THE SOCIETY FOR APPLIED BACTERIOLOGY
TECHNICAL SERIES No. 23

Industrial Microbiological Testing

Edited by

J. W. HOPTON
Department of Microbiology
The University of Birmingham, UK

E. C. HILL
E. C. Hill & Associates
Cardiff

BLACKWELL SCIENTIFIC PUBLICATIONS
OXFORD LONDON EDINBURGH
BOSTON PALO ALTO MELBOURNE

Blackwell Scientific Publications
Editorial offices:
Osney Mead, Oxford OX2 0EL
(*Orders*: Tel. 0865 240201)
8 John Street, London WC1N 2ES
23 Ainslie Place, Edinburgh EH3 6AJ
52 Beacon Street, Boston
Massachusetts 02108, USA
667 Lytton Avenue, Palo Alto
California 94301, USA
107 Barry Street, Carlton
Victoria 3053, Australia

First published 1987

DISTRIBUTORS

USA and Canada
Blackwell Scientific Publications Inc
PO Box 50009, Palo Alto
California 94303
(*Orders*: Tel. (415) 965-4081)
Australia
Blackwell Scientific Publications
(Australia) Pty Ltd
107 Barry Street,
Carlton, Victoria 3053
(*Orders*: Tel. (03) 347 0300)

Typeset, printed and bound in Great Britain by
William Clowes Limited, Beccles and London

British Library
Cataloguing in Publication Data
Industrial microbiological testing.—
(The Society for Applied Bacteriology
technical series; v. 23).
1. Biotechnology—Technique
I. Hopton, J. W. II. Hill, E. C.
660'.62'028 QR53

ISBN 0–632–01793–7
ISSN 0–583–8924

Library of Congress
Cataloging-in-Publication Data
Industrial microbiological testing.
(Technical series/the Society for Applied
Bacteriology; no. 23)
"The descriptive texts of contributions made to the Demonstration Meeting on Industrial Microbiology Testing, held at the Thames Polytechnic on 16 October 1985"—Pref.
1. Industrial microbiology—Congresses.
2. Micro-organisms—Identification—Congresses. 3. Contamination (Technology)—Congresses. 4. Pesticides—Testing—Congresses. I. Hopton, Jack W. II. Hill, E. C. III. Demonstration Meeting on Industrial Microbiology Testing (1985: Thames Polytechnic) IV. Series: Technical series (Society for Applied Bacteriology); no. 23.
QR53.I518 1987 576'.16 87–10294

ISBN 0–632–01793–7

A complete list of titles in the Society for Applied Bacteriology Technical Series appears at the end of this volume.

Contents

Contributors

D. Allsopp, *Commonwealth Mycological Institute, Ferry Lane, Kew, Surrey TW9 3AF*

J. Ashworth, *Albury Laboratories Ltd, The Old Mill, Albury, Guildford, Surrey GU5 9AZ*

P. G. H. Bakker, *IWS, Valley Drive, Ilkley, LS29 8PB*

Stefania Bobowski, *Biological Sciences Branch, BP Research Centre, Sunbury on Thames, Middlesex TW16 7LN*

I. J. Bousfield, *National Collections of Industrial and Marine Bacteria Ltd, Torry Research Station, PO Box 31, 135 Abbey Road, Aberdeen AB9 8DG*

A. F. Bravery, *Biodeterioration Section, Princes Risborough Laboratory, Building Research Establishment, Princes Risborough, Aylesbury, Buckinghamshire HP17 9PX*

I. J. Bruce, *School of Biological Sciences and Environmental Health, Thames Polytechnic, Wellington Street, London SE18 6PF*

Jennifer S. Colbourne, *Thames Water, New River Head, 177 Rosebery Avenue, London EC1R 4TP*

R. A. Collett, *Worcester College of Higher Education, Worcester WR2 6AJ*

I. H. Evans, *School of Biological Sciences and Environmental Health, Thames Polytechnic, Wellington Street, London SE18 6PF*

K. P. Flint, *Department of Biological Sciences, University of Warwick, Coventry CV4 7AL*

P. D. Gilbert, *Shell Research Centre, Sittingbourne, Kent ME9 8AG*

P. N. Green, *National Collections of Industrial and Marine Bacteria Ltd, Torry Research Station, PO Box 31, 135 Abbey Road, Aberdeen AB9 8DG*

P. F. Hamlyn, *The Shirley Institute, Wilmslow Road, Didsbury, Manchester M20 8RX*

Lesley Harrison, *Department of Microbiology, The Medical School, The University, Newcastle upon Tyne NE2 4HH*

B. N. Herbert, *Shell Research Centre, Sittingbourne, Kent ME9 8AG*

R. A. Herbert, *Department of Biological Sciences, The University, Dundee DD1 4HN*

E. C. Hill, *E. C. Hill & Associates, Unit M22, Cardiff Workshops, Lewis Road, East Moors, Cardiff CF1 5EG*

G. C. Hill, *E. C. Hill & Associates, Unit M22, Cardiff Workshops, Lewis Road, East Moors, Cardiff CF1 5EG*

J. W. Hopton, *Department of Microbiology, The University, Birmingham B15 2TT*

Joan Kelley, *Commonwealth Mycological Institute, Ferry Lane, Kew, Surrey TW9 3AF*

Pamela McAthey, *Polytechnic of North London, Holloway, London N7 8DB*

B. J. MacCarthy, *Wira Technology Group, Wira House, West Park Ring Road, Leeds LS16 6QL*

J. M. Maw, *School of Natural Sciences, Hatfield Polytechnic, College Lane, Hatfield, Hertfordshire AL10 9AB*

L. H. G. Morton, *Faculty of Science, Lancashire Polytechnic, Preston, Lancashire PR1 2TQ*

D. B. Nedwell, *Department of Biology, University of Essex, Wivenhoe Park, Colchester CO4 3SQ*

P. D. Oates, *Castrol Research Laboratories, Pangbourne, Berkshire*

Bharti Patel, *Polytechnic of North London, Holloway, London N7 8DB*

D. Powell, *Admiralty Research Establishment, Dockyard Laboratory, Portsmouth*

A. P. Rawlinson, *Biological Research Branch, BP Research Centre, Sunbury on Thames, Middlesex TW16 7LN*

Elaine A. Rees, *Department of Biological Sciences, The University, Dundee DD1 4HN*

D. A. Robbins, *E. C. Hill & Associates, Unit M22, Cardiff Workshops, Lewis Road, East Moors, Cardiff CF1 5EG*

Moya A. Russ, *Department of Biological Sciences, The University, Dundee DD1 4HN*

B. F. Sagar, *The Shirley Institute, Wilmslow Road, Didsbury, Manchester M20 8RX*

P. S. Senior, *Wira Technology Group, Wira House, West Park Ring Road, Leeds LS16 6QL*

Jean L. Shennan, *BP Research Centre, Sunbury on Thames, Middlesex TW16 7LN*

R. N. Smith, *School of Natural Sciences, Hatfield Polytechnic, College Lane, Hatfield, Hertfordshire AL10 9AB*

Angela Stones, *Durham Chemicals Ltd, Birtley, Chester-Le-Street, Co. Durham*

Caroline S. Tughan, *Department of Biological Sciences, The University, Dundee DD1 4HN*

D. S. Wales, *The Shirley Institute, Wilmslow Road, Didsbury, Manchester M20 8RX*

R. Whittle, *Vosper-Thornycroft (UK) Ltd, Southampton*

D. Wilkie, *Department of Botany and Microbiology, University College, Gower Street, London WC1E 6BT*

Preface

This volume, number 23 in the Society for Applied Bacteriology Technical Series, contains the descriptive texts of contributions made to the Demonstration Meeting on Industrial Microbiology Testing, held at the Thames Polytechnic on 16 October 1985.

The editors were members of both the Society for Applied Bacteriology and the Institute of Petroleum Microbiology Group and the venture was the outcome of an expressed desire to hold a joint meeting which would appeal to members of both organizations.

The programme was compiled by the editors, and the local arrangements at Thames Polytechnic were in the hands of Dr B. N. Kliger and his staff.

The editors wish to express their thanks to all those colleagues who were involved in the organization and administration of the meeting and it is hoped that the finished volume will justify the effort which the contributors willingly gave to the production and exhibition of their demonstrations and the preparation of their papers. The editors are especially grateful to Dr Susan M. Passmore for her guidance and help.

J W Hopton
E C Hill

Methods for Assessing the Efficacy of Fungicides for Wood and Wood Products

A. F. BRAVERY
Biodeterioration Section, Princes Risborough Laboratory, Building Research Establishment, Princes Risborough, Aylesbury, Bucks.

The principles of laboratory testing of wood preservatives against fungi and the main factors which dictate the development of the methods were reviewed in a previous paper (Bravery 1975).

The effectiveness of a wood preservative in any particular timber depends upon: (a) the toxicity and permanence of the preservative itself, (b) the method of application, and (c) the environment in which the treated wood is exposed (Smith 1965; Gibson 1966). Methods for assessing effectiveness must therefore take these factors into account and it is immediately clear that an evaluation relevant to all three requires several different tests.

Factors Affecting the Development and Selection of Test Methods

Environment of Use and Types of Fungi

Many different fungi cause deterioration of wood and wood products infecting the timber at different stages from initial felling of the tree through to any one of the wide range of end-uses. The fungi can be grouped on the basis of similarities in their effects on the wood which in turn reflect different physiological capabilities and nutritional requirements; these groups are sapstain, mould, blue stain in service, soft rot and true decay inducers. This is not to imply that the groupings are invariable because some fungi can display characteristics of more than one group depending upon the species of wood and the specific environmental conditions prevailing (Levy 1967). However, the groupings form the basis against which test methods for determining the intrinsic toxicity of fungicides have been developed.

Arising from their different physiology and nutritional requirements different fungi induce different sorts of defects in wood and wood products. Hence fungicides may be required to prevent spoilage, disfigurement or degradation according to the

Industrial
Microbiological Testing

0-632-01793-7

type of fungal invasion and the end-use of the commodity. Furthermore, apart from preventing initiation of attack of wood, fungicides are sometimes used to eradicate existing fungal infection, in which case different levels of activity and methods of presentation of the preservatives may be involved.

Wood and Preservative Distribution

Wood preservatives can be applied by brush, spray, immersion, diffusion or a variety of impregnation techniques using industrial pressure and vacuum plant. Furthermore, different timber species vary, sometimes markedly, in their ability to accept fluids, i.e. their treatability. Because of this and the variety of methods of application, the distribution of the preservatives in wood can be markedly different. Since effectiveness depends in part upon the extent of distribution of the preservative within the wood this factor must be capable of accommodation in the methods of assessment of efficacy.

The distribution of a preservative in timber is influenced not only by the treatability of the wood species and the method of application but also by the characteristics of the fluid itself. Thus whilst both water-based and organic solvent fluids can be impregnated deeply into dry permeable timber using vacuum/pressure techniques, water-based systems generally penetrate much less than do organic solvent systems when applied by simple brush, spray or immersion methods. In contrast, simple diffusion techniques can provide deep penetration into green (unseasoned) timber.

Further special factors apply to wood products, represented mainly by composite sheet materials. These often have particular properties arising from the mixtures of wood species and the manufacturing process involved in their production, the size, shape and configuration of the particles or laminae, and the adhesives or other bonding agents or additives. These properties in turn influence preservative distribution, choice of test fungi and test conditions.

Categories of Test

There are two main groups of methods for assessing the efficacy of fungicides for wood and wood products, namely laboratory tests and commodity type end-use field tests. Laboratory tests can differ markedly employing defined pure cultures, mixtures of pure cultures or the natural microflora in procedures designed to control and optimize conditions and to accelerate rates of fungal attack. Commodity/field tests employ samples representative of particular treated wooden components exposed under conditions representative of those expected in service. Field tests require many years to carry out, whereas most laboratory tests are designed to give results within 6–16 weeks.

Data obtained from laboratory tests do not lend themselves directly to prediction of actual treatment levels or performance life of a commodity in service. They allow for determination of the basic toxicity of a fungicide before and after any specific preconditioning treatment such as leaching or evaporative ageing. By determining the toxicity levels for new or experimental materials in comparison with those for reference materials of known performance in any given end-use, an estimate of likely service life or equivalent treating level can be derived. However, even data from field tests must be interpreted with caution since it is often not possible to simulate in a field test the full range of factors which will influence the behaviour of fungicide in all conceivable end-uses. In the main, field trials aim to maximize the severity of the exposure relative to a particular commodity end-use so as to provide a severe yet realistic challenge to the fungicide. Samples for field tests are of a realistic size and orientation and so permit application of the fungicidal solution by a realistic method of treatment. In laboratory tests the samples in general must be small enough to allow insertion into closed containers and to facilitate rapid changes in weight or strength by the fungi. As a consequence the samples often are too small to permit realistic methods of application so that penetration is complete and even throughout without the normally expected concentration gradient from surface to interior.

The range of laboratory tests available for the bioassay of fungicides for wood and wood products were presented diagrammatically in a previous paper (Bravery 1975). That diagram still provides a valid basis for rationalizing the various types of test but needs modification to take into account developments which have happened since (Fig. 1).

Laboratory Tests Employing Bioassay Plates

Agar Plate Test

Attempts to develop and standardize an agar plate bioassay test have largely been abandoned. The reliability of these methods has been questioned from a very early stage (Rabanus 1931; Finholt *et al.* 1952; Smith & Savory 1965; Dickinson 1974; Bravery & Carey 1977). Discrepancies in results between agar plate tests and wood block tests have been variously attributed to differences in the extracellular enzyme secretions of the test fungi under different test conditions, to the absence in agar tests of preservative/wood interactions and to difficulties in obtaining homogeneous dispersions of oil-soluble chemicals within the agar substrate.

A development of plate tests incorporating fungicides, involves assessment of toxicity by measurement of zones of inhibition in agar plates (Fig. 2). Suppression of growth of the indicator organism is measured and quantification of the dose response made possible by reference to a calibration curve. Scheffer & Graham

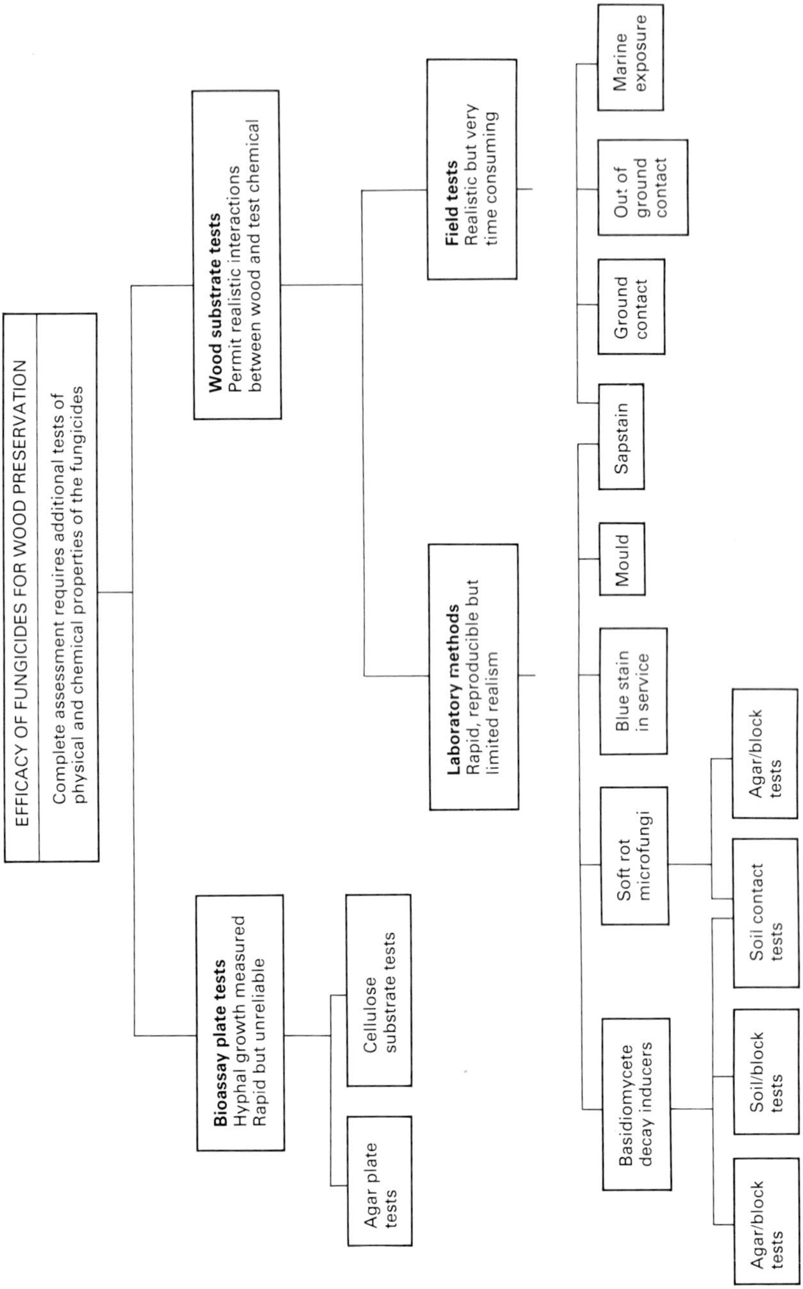

FIG. 1. Methods for determining the efficacy of fungicides for wood preservatives.

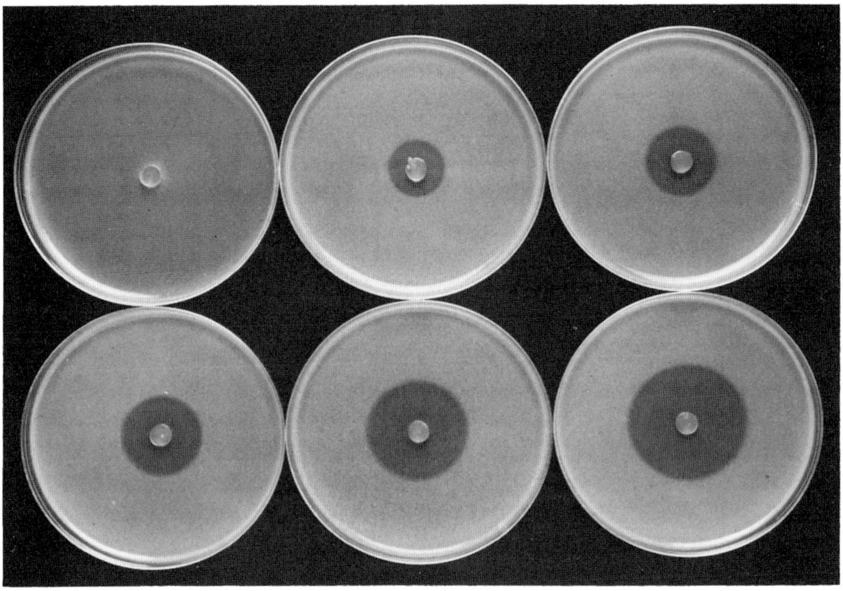

FIG. 2. Agar plate bioassay using zones of inhibition of *Bacillus subtilis*.

(1973) and Scheffer & Lew (1976) proposed a technique using the wood destroying basidiomycete *Poria placenta* as the indicator organism, and Scheffer & Gollob (1978) described a variation of this procedure in which *Aspergillus niger* was used as the assay organism. Sutter & Carey (1985) have proposed a further variant based on the widely accepted agar diffusion technique for antibiotics described by Drews (1976) in which *Bacillus subtilis* is the indicator organism. All of these methods depend upon the fungicide having a measure of solubility and hence diffusibility in an agar gel. Anomalies in quantification can arise too if the toxicants are volatile and cause inhibition through aerial concentrations. Their relevance in predicting potential performance in wood depends upon relationship to known reference preservatives and even then the other limitations of agar-based techniques, such as the absence of provision for interaction with a woody substrate, still apply.

Cellulose Substrate Tests

In attempts to overcome the limitations of agar tests, adaptations of the bioassay plate approach have been developed in which sawdust, wood flour, wood pulp and paper or card discs have been used as the test medium (Dickinson 1974; Bravery & Carey 1977; Greaves 1977; Thornton 1977). Dickinson (1974) achieved results with filter paper discs which compared very favourably with those from wood block

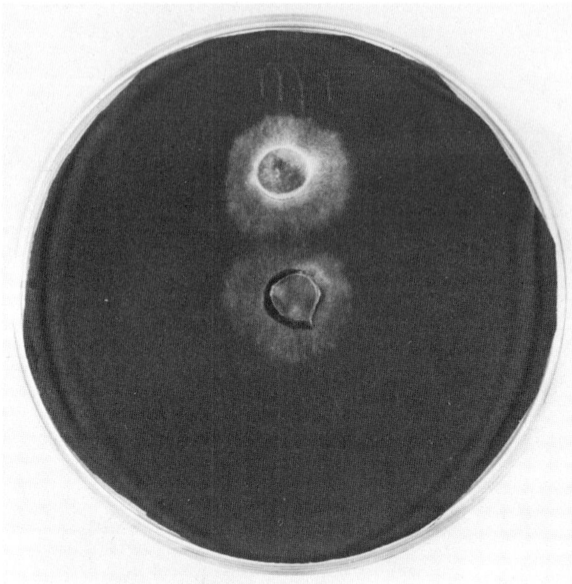

FIG. 3. Cellulose disc bioassay test.

tests and Bravery & Carey (1977) identified several alternative cellulosic substrates and defined conditions for their use. The cellulose substrate bioassay (Fig. 3) is now established as a potentially useful technique for rapid screening of fungicides as wood preservatives and for determining possible reductions in efficacy by evaporation, leaching or formulation changes. However, it is only suitable for those fungi able to grow on cellulose substrate.

Laboratory Tests Employing Wood Substrates

Tests for Control of Microfungi

Sapstain fungi

Sapstain fungi are of economic importance in causing blue-black and brown disfigurement of freshly felled logs or sawn timber. They only infect timber so long as it is green, and are controlled by rapid drying. Because handling practices sometimes cannot ensure rapid surface drying, chemical treatments are used but need only be effective for the comparatively short period until the timber dries. Products and components manufactured from wood are not vulnerable to sapstain fungi.

A variety of different laboratory tests have been proposed to assess the potential of fungicides for controlling sapstain (Fig. 4). Hatfield *et al.* (1950) proposed the

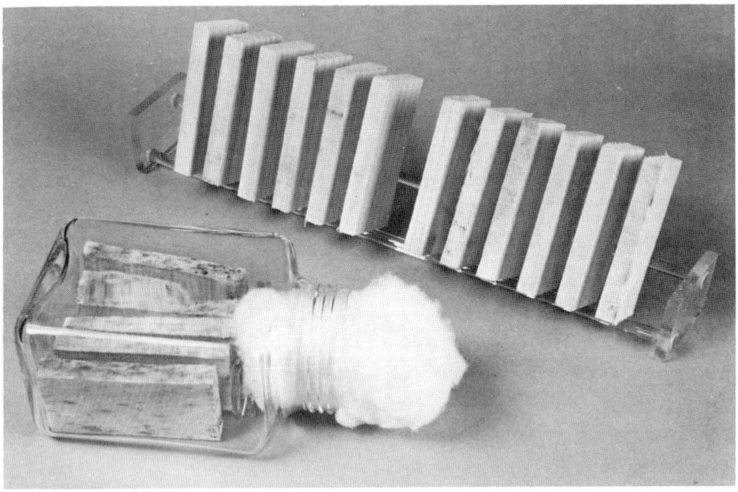

FIG. 4. Miniature board test for sapstain preventives.

unseasoned wood disc method, Schulze (1951) a small, impregnated wood cube test over filter papers and Cserjesi & Roff (1970) developed a technique using miniature wood boards incubated over damp filter papers. In these and other similar tests a mixed inoculum of pure cultures is used to infect the test pieces. Such laboratory tests are still practised for screening work (Cserjesi & Johnson 1982; Eslyn & Cassens 1983; Lewis *et al.* 1985). However, the advantages of using miniaturized samples in laboratory tests have been questioned by some workers who have chosen to use larger samples and conditions of infection and storage more representative of those likely to be encountered in practice (Butcher 1973; Hulme & Thomas 1979; Cserjesi 1980; Drysdale & Preston 1982). Indeed, conclusions from discussions of a special sub-committee within the International Research Group on Wood Preservation were that speed and reliability of results were maximized by initiating trials immediately in the field. This has been the pattern of many recent studies (Dickinson & Henningsson 1984; Lewis *et al.* 1985) in which fresh-cut boards of a size close to commercial practice are dip-treated with test formulations on location in the field and stacked close or open-piled, sheeted or unsheeted, as appropriate. Reliance on natural infection further ensures that the troublesome more resistant local mould species are available as part of the natural microflora together with the recognized sapstain species of *Ceratocystis, Leptographium, Graphium* and *Phialophora*. Visual assessments are made of extent of blue-black or other disfigurement as a criterion of success or failure of the fungicidal treatments.

Mould fungi

Chemical treatments are not normally applied solely and specifically to prevent infection of timber by mould fungi. Exceptions to this generality are mould species mixed with sapstain fungi as discussed above and occasional mould infections which sometimes occur immediately after treatment of timber with some water-borne preservatives. Indeed, mould fungi are generally symptomatic of damp environmental conditions which are in themselves detrimental, and control is best effected by providing drying conditions. Some types of wood products may become infected in service, however, and they require surface sterilization.

Test methods available include miniature board tests of the types described for sapstain fungi, and there are also standardized procedures in British Standard 1982:1968 which was issued in a revised form in 1986. Two types of test are provided for, one at high humidity but low moisture content and one at high humidity but also under sustained conditions of high moisture content. In the former, samples are placed in open petri dishes within a closed humidity chamber containing moist Vermiculite. In the latter test, samples are placed in direct contact with a mineral salts agar medium in open petri dishes within the same type of closed humidity chamber. An alternative variation on the high humidity/high moisture content test has been proposed for evaluation of chipboards (Bravery 1975) in which samples are exposed on beds of moist sawdust.

Blue stain in service

This defect is caused by specific fungi which are different from the main species causing sapstain, though they do have pigmented hyphae. Fungi causing blue stain in service mainly colonize the surfaces of timber in service through and beneath paints and finishes (Dickinson 1971; Bravery & Miller 1980). The earliest recognition of the need for specific tests for fungicides against blue stain in service can be attributed to Schulze (1952). He proposed a test using small boards (10 × 50 × 9 mm) of planed softwood treated with fungicidal primers and inoculated with the particular fungi which cause blue-black or brown disfigurement of seasoned timber in service. Butin (1961) improved the technique by specifying infection with two particular fungi *Pullularia (Aureobasidium) pullulans* and *Schlerophoma pithyophila*. Dickinson (1971) developed a simpler and more rapid screening test in which pairs of small treated sapwood veneers (100 × 25 mm) were bonded to an aluminium laminate, artificially weathered and sprayed with a mixed spore suspension and incubated in a humidity cabinet.

The significance of fungi causing bluestain in service is now well established (Bravery & Miller 1980) and the need for methods to assess the efficacy of fungicides has been recognized in the production of a draft European Standard Method (Comité Européen de Normalisation 1985a; Fig. 5). One part of the method is based on the technique of Butin (1961) and relates to products applied by surface applications. The second part relates to deeper penetrating treatments and has been derived from work in the U.K. (Bravery & Derbyshire 1979) subsequently refined by co-operative international effort (Bravery & Dickinson 1985). Both methods

FIG. 5. European Standard Method of test of chemicals to control blue stain in service.

involve assessment of the extent to which the specific fungi proposed by Butin (1961) are able to disfigure the surface of treated test panels after a period of either natural or artificial weathering. In each case provision is made for test samples which can receive a realistic treatment by brush, dipping or even vacuum pressure techniques prior to weathering and exposure to the test fungi under controlled conditions for 6 weeks.

Soft rot fungi

Soft rot of timber is caused by cellulose-destroying microfungi belonging to the Ascomycetes and Fungi Imperfecti (Savory 1954a, b). They are common in soil and water and colonize wood out of ground contact when moisture conditions are too wet or too variable to permit active development of Basidiomycete decay fungi. Although individual fungi can be isolated and shown in pure culture to cause soft rot under natural conditions, soft rot decay results from the action of a complex of different micro-organisms including bacteria and actinomycetes. Selection of suitable species for laboratory testing is complicated because ecological studies have not identified consistently any particular causal species and because few of those isolated are sufficiently active in pure culture for laboratory test purposes. Though *Chaetomium globosum* has become widely accepted as a convenient test organism at least for hardwoods, its precise practical relevance remains uncertain.

Because of these difficulties, two classes of soft rot tests have evolved. One uses pure cultures on a mineral salts agar medium (Fig. 6) and the other involves burial of test blocks in soil or vermiculite substrates relying either on the natural microflora

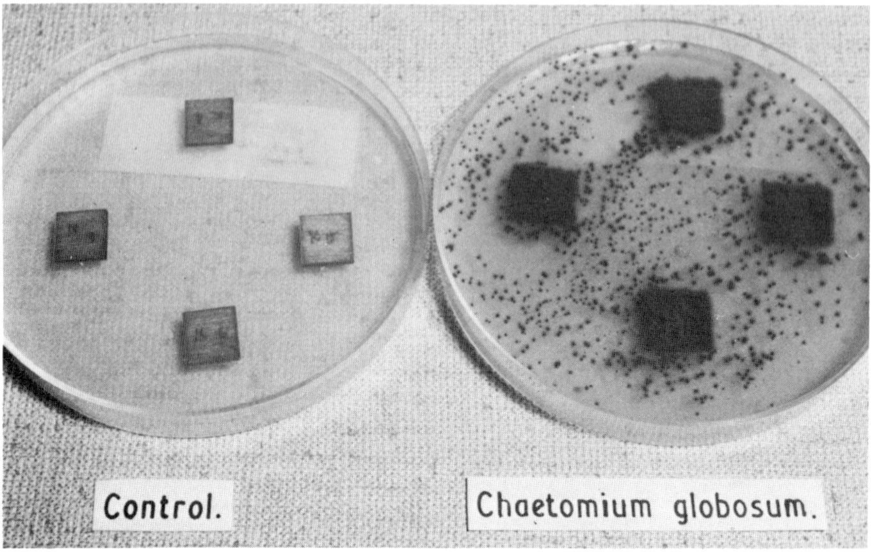

FIG. 6. Petri dish agar/wood block bioassay against soft rot fungi.

of the soil or inoculation with pure or mixed cultures or 'seeding' with the natural microflora from unsterile soil. Methods for determining the effectiveness of wood preservatives against soft rot fungi were reviewed by Savory & Bravery (1970a,b). More recently, Gersonde & Kerner-Gang (1976) have reviewed information relating to improvement in the reproducibility of burial systems. Pure culture agar tests have the advantage of being better defined than soil tests but the choice of fungal species to obtain vigorous attack in softwoods poses problems. Whilst natural soil burial can provide greater aggressiveness, definition and reproducibility of the soil then presents difficulties.

Although an improved, controlled laboratory test has been developed using burial of blocks in Vermiculite (Kerner-Gang 1974, 1975), such techniques have still not gained general acceptance. Indeed, there has been a growing preference internationally for techniques based on burial of small blocks or miniature stakes in beds of natural soil (Bravery 1968; Butcher 1983; Hedley 1983). Whilst standardization of soil burial methods for testing fungicides has not yet proved possible because of the variability of the soils, the system can be used successfully on a local basis employing preferred soils of known aggressiveness (Drysdale 1984). Small stakes, 100 × 10 × 5 mm or larger in size, are inserted vertically so as to leave a quarter to a third exposed above the soil surface (Fig. 7). The soil beds, which may be in small bins or large tanks, are kept moist and are monitored continuously for aggressiveness by insertion of reference stakes of particular wood species impregnated with particular preservatives.

Criteria for measurement of effectiveness of fungicides may involve loss in strength or weight. In tests using small test blocks, weight loss has been preferred with the accepted minimum significant loss accepted generally as 3%. Measurements of loss in strength have also been used in attempts to develop a more sensitive test (Mateus 1954; Hartley 1958; Wälchli 1969; Smith 1971; Vinden *et al.* 1982, 1983). In soil bed tests the strength criterion can be used at its very simplest by a manual bending test. This depends upon the operator's own skill in applying a broadly consistent bending force, and recording the time interval at which each sample in turn fails; a 'mean service life' can then be derived. In general, the weight loss criterion is preferred, however, because it is objective, and despite being destructive is simple to carry out and requires unsophisticated equipment or sample preparation. Indeed, this criterion has been adopted together with the soil bed technique in the current revision of BS 1982 relating to assessment of the resistance to cellulose decomposing microfungi (soft rot) of wood-based board materials.

FIG. 7. Mini stake/soil bed test against natural soft rot microflora.

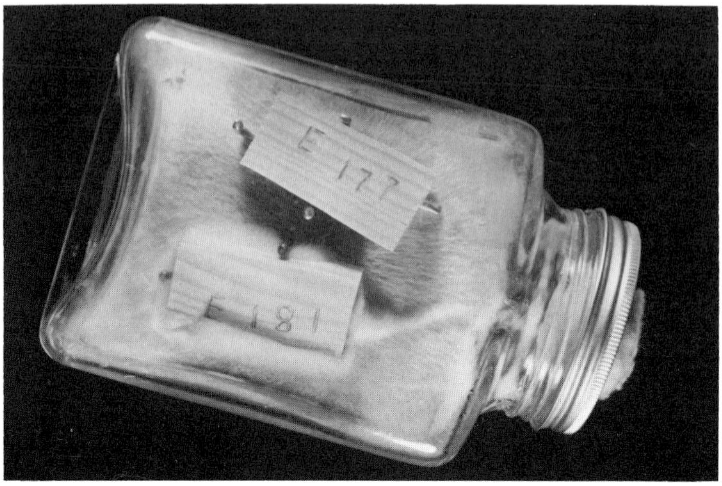

FIG. 8. European Standard agar/wood block test against Basidiomycete decay fungi.

Tests for Control of Wood-Destroying Basidiomycete Fungi

Toxic values tests

Within Europe, including Scandinavia, the Standard Method adopted for determining the efficacy of fungicides against wood-attacking Basidiomycetes is as described in BS 6009:1982–European Standard EN 113 (British Standards Institution 1982). The principle of the method is that small blocks of pine sapwood (50 × 25 × 15 mm) are impregnated with a graded series of dilutions of the fungicidal product under test and exposed in jars over malt agar to pure cultures of selected test fungi (Fig. 8). Determination of losses in oven dry weight over the 16-week incubation period allow derivation of the so-called toxic values, being the highest concentration which just permits attack, and the lowest which just prevents it. The technique is intended to provide toxic values for a fungicide before and indeed after standardized preconditioning to assess loss by evaporation or leaching using either or both of BS 5761 : Part 1 : (BSI, 1979) or BS 5761 : Part 2 (BSI, 1980). It is not a method intended to enable assessment of the efficacy of a preservative treatment applied by any particular method of application. The European Standard Method supersedes earlier, French, English and Nordic methods. In North America the Standard Method D-1413-1976 (American Society for Testing Materials, 1976) is a soil/block technique based on that devised by Leutritz (1946) in which small wood cubes (19 × 19 × 19 mm) treated with the preservative are placed on thin untreated wood ‘feeder’ blocks over moist sterile soil which is inoculated with a pure culture of the particular test fungi. Essentially the same technique though sometimes with modifications has been favoured in New Zealand also (Hedley & Foster 1972; Butcher *et al.* 1977a,b). A soil/block method offers advantages over the agar/block system preferred in Europe in that moisture conditions can be maintained more consistently; any potentially inhibitory volatile constituents of the preservative treatment can be adsorbed and the feeder strip is a more natural source of infection. On the other hand, soil is more inconvenient and time consuming to prepare and variations in its source can influence results. In general both methods are capable of giving reliable and comparable toxic values. A variation on the soil/block test using quartz sand is employed as a Japanese Standard Test JIS A 9302 (Hedley *et al.* 1982).

Agar/block and soil/block methods of the types described above have the disadvantages of requiring comparatively long incubation periods as well as being comparatively bulky tests. Improvements in speed and simplicity have been sought by various workers, mainly by reducing the size of the test block and of the culture container (Bravery 1978; Richardson 1978; Cox 1981; Bravery 1983; Imsgard *et al.* 1985). Earlier attempts to employ respirometric methods (Smith 1967) for routine testing have largely been abandoned.

Surface treatments tests

A variety of techniques have been developed for giving indications of the efficacy against Basidiomycetes of fungicides applied by surface treatments. Savory & Carey (1976) adopted the traditional agar/block and soil/block type of test to assess treatments for wooden joinery, whilst others have developed entirely new approaches. Frequently these are devices for bioassaying depth of penetration as much as toxicity (Sutter 1978; Ingleby 1982; Saunders & Blow 1985). Essentially the methods involve focusing the decay fungus on to a treated surface and assessing its capability of penetrating into and through the treatment to reach untreated wood within. In the case of Sutter's test, cross-cut 'biscuits' are placed on cultures so that the fungus has access to the untreated core whilst resistance of the fungicidal treatment in the outer zones is detected by difference.

The principle of all the methods so far referred to for Basidiomycetes is that they are very severe tests, the conditions for the fungus being intentionally optimized. Various workers have attempted to moderate the degree of severity or to develop novel approaches by tailoring the techniques to particular end-use requirements. Examples are the polythene bag technique of Amburgey & Behr (1979), the so-called 'above-ground test' developed in New Zealand (Butcher 1979; Hedley *et al.* 1982) and innovative small-scale assay techniques from Australia for studying fungicidal effects against *Serpula lacrymans* (Thornton & Collett, 1979; Thornton, 1979).

Tests Employing Wood-Based Boards

Methodology for assessing fungicidal additives or treatments applied to wood-based boards has been developing rapidly in recent years, as the range of board types has grown and the extent to which they are used in risk situations has increased. A recent review of some of the methods, the philosophy behind them and approaches to application of preservatives has been published by Hedley (1982). The revision of BS 1982 issued in 1986 presents the methodology currently preferred in the U.K. together with guidance in its use. Laboratory tests in the main are developed from those established for solid wood with adaptations to suit particular special needs (Savory 1969; Jackson & Savory 1968; Lea 1982; Hedley 1976). For example, when sheet materials are cut into small pieces the proportion of edge grain is unrealistically high and this is more vulnerable to fungal invasion. Techniques for sealing edges have therefore been developed. Inhibitory volatiles from adhesives must be dispersed before test and fungal strains more relevant to service experience take priority; in particular the white rot *Pleurotus ostreatus* has been introduced as a specific test fungus for board materials (British Standards Institution 1986).

Alternative approaches have been developed to obviate the potential anomalies of cutting small samples to place inside culture vessels. Exposure of large sample pieces to vigorous pure cultures in a fungus cellar has been used (Deppe & Gersonde 1977), whilst others have localized fungal attack on sample sheets using concentrated inocula of culture plates or already infected wood (Taylor 1978; Trong 1984; Deon & Trong 1985).

Tests Involving in-situ Fungicides

A specialized field of assessment of fungicides for wood preservation is that concerned with *in-situ* treatment against decay in service, of both the Basidiomycete and soft rot type. Usually the two types of decay occur together and must be treated together, and the main commodities which are the subject of treatment include poles, posts and railway sleepers (Greaves 1977; Greaves *et al.* 1982; Beckgaard *et al.* 1980), timber structures such as decking, bridges, wharfs, jetties (Highley 1980, 1984; Highley & Eslyn 1982; Helsung *et al.* 1984) and timber building components, especially joinery (Henshaw 1978; Lea 1980; Dicker *et al.* 1983; Blow & Summers 1985). Approaches to assessment almost always involve a combination of techniques. Included are novel, non-standard laboratory tests in which fungicides are applied to sawdust or small blocks of wood in ways similar to those described elsewhere in this paper. More important and relevant are field trials or simulated trials in which the fungicides are applied to timber in service situations and bioassay sampling techniques are used to determine residual viability of the decay inducing fungi (as an indication of sterilising efficacy) and residual preventive efficacy.

Field tests

Ground Contact Tests

The traditional and ultimately most conclusive demonstration of the capability of any fungicide to prevent decay is successful performance under field exposure conditions for prolonged periods of time. Ground contact tests have been established by different Research Institutions world-wide, and results of 50 years of such trials at Princes Risborough were reviewed by Purslow (1979). The ground contact trial consists of placing in the ground, stakes of a particular size, cut from a particular wood species, treated with a particular preservative to a particular specification (Fig. 9). The stakes are examined visually and tested for soundness, usually annually, and a mean service life derived. A Standard Method was adopted by the International Union of Forest Research Organisations (Becker 1972) and a system based on this is due to be standardized within Europe in 1986 (Comité Européen de Normalisation 1985b). Reduction of the dimensions of the test samples can produce results more rapidly and a large-scale co-operative test using smaller samples has been established within the International Research Group on Wood Preservation (Dickinson 1976; Dickinson & Levy 1982).

The principal disadvantage of the so-called 'graveyard test' approach to assessing

FIG. 9. 'Graveyard' field test for wood preservative fungicides.

efficacy is that with effective products it can take many years before any indication of performance can be obtained. Attempts to accelerate the process by bringing the field into the laboratory are being investigated. Large-scale beds of natural soil are established under constant conditions in which it is intended to encourage Basidiomycetes as well as soft rot fungi (Johnson *et al.* 1982).

Out-of-Ground-Contact Tests

With the advent in the late 1960s of specifications in the U.K. requiring pretreatment of softwood window joinery as well as other commodities such as cladding, there has been a need for field exposure tests relevant to this type of end use. Work at Princes Risborough Laboratory (Banks 1971; Purslow & Williams 1978; Purslow 1979) pioneered the use in the U.K. of the L and T joint systems which had their origins in the U.S.A. (Lance 1958; Verrall 1961). In essence L or T shaped mortice and tenon joints are prepared in sizes close to commercial practice and after application of an appropriate preservative treatment they are coated, or left uncoated, and placed on racks designed to encourage entrapment of water in the mortice and tenon (Fig. 10). Regular visual assessments of the condition of the samples and in particular of the development and extent of decay enable derivation of service lives. However, as with ground contact tests, several years must pass before data which can be used for assessment of performance can be obtained. Work by Savory *et al.*

FIG. 10. Out-of-ground-contact field test for wood preservative fungicides.

(1977), Savory & Carey (1979), Carey (1980), Carey (1982) and Carey & Bravery (1985) has established a basis for assessing performance during the first $\frac{1}{2}$–3 years by destructive analysis of small L-joints. Comparison of measurements of increases in porosity of the timber and of the time when Basidiomycete fungi are first detected gives indications of the relative performance of a fungicide or treatment compared with the performance of known reference chemicals.

The L-joint trials also permit assessment of the efficacy of fungicides in the preservative pretreatments or indeed in the coatings applied, to resist colonization by mould and fungi causing blue stain in service. However, the more conventional field approach to assessing efficacy against these fungi is the panel test (Boxall 1981, 1984) in which incidence of fungi is one of a number of defects noted and recorded at intervals on coated panels exposed to the weather.

Marine Exposure Tests

Assessment of fungicides against marine fungi causing decay is not normally conducted routinely using laboratory tests because of the difficulties in providing conditions which appropriately model the marine environment. As a consequence, such information is more usually gained as part of an overall assessment from 'field tests' involving panels exposed on rigs suspended from rafts or attached to piers, etc., in the sea. In most situations the main physical deterioration in such tests is likely to be induced by marine borers of a particular type depending upon the geographical location of the trial site. A new European Standard Method, published in 1986 (Comité Européen de Normalisation, 1984), defines the procedure adopted within Europe.

Conclusions

Assessment of the efficacy of fungicides for the preservation of wood and wood products requires a wide range of tests. Some of these can be conducted comparatively quickly under reproducible and controlled conditions in the laboratory, but others necessarily entail outdoor exposure to natural populations of fungi and other micro-organisms. The groups of fungi requiring specific test procedures include sapstain fungi, moulds, fungi causing blue stain in service, soft rot microfungi and decay-inducing Basidiomycetes.

Simplified bioassay plate tests give data of limited value and tests involving the treatment of wood blocks and exposure to pure or mixed cultures of active fungi are necessary. Simple visual criteria or measurements of weight loss are normally used though strength loss is convenient and acceptable for some types of test.

Ultimately field or simulated field exposure is necessary, and much research work is currently being devoted to improvements in these methods so as to provide data more quickly.

References

AMBURGEY, T. L. & BEHR, E. A. 1979 Accelerated method for testing wood preservatives for above ground use. *Materials and Organisms* **14** (2), 141–152.

AMERICAN SOCIETY FOR TESTING AND MATERIALS 1976 Standard method of testing wood preservatives by laboratory soil block cultures. ASTM American National Standard D1413–76.

BANKS, W. B. 1971 The rôle of water repellents in the protection of timber. *Records of the Annual Convention of the British Wood Preserving Association*, pp. 129–147.

BECKER, G. 1972 Suggested standard method for field tests with wooden stakes *PANS* **18**, (1), 137–142.

BECKGAARD, C., LENNART, B., HENNINGSSON, B. & JERMER, J. 1980 Remedial treatment of creosoted railway sleepers of redwood by selective application of boric acid. International Research Group on Wood Preservation Document No. IRG/WP/3134.

BLOW, D. P. & SUMMERS, N. A. 1985. A laboratory evaluation of fused borate rods for the treatment of timber. *International Biodeterioration* **21**, (1), 27–40.

BOXALL, J. 1981 Coating formulation; its effect on the durability of exterior wood finishes. *Journal of Oil and Colour Chemists Association* **64**, 135–139.

BOXALL, J. 1984 Exterior wood finishes: performance testing by accelerated natural weathering. *Journal of Oil and Colour Chemists Association* **2**, 40–44.

BRAVERY, A. F. 1968 Determining the tolerance of soft-rot fungi to wood preservatives: a comparison of test methods. *Materials and Organisms* **3**, (3), 213–217.

BRAVERY, A. F. 1975 Microbiological assay of chemicals for the protection of wood. In: *Some Methods for Microbiological Assay.* Technical Series No. 8, pp. 203–221. London: Academic Press.

BRAVERY, A. F. 1978 A miniaturised wood block test for the rapid evaluation of wood preservative fungicides. International Research Group on Wood Preservation Document No. IRG/WP/2113.

BRAVERY, A. F. 1983 Report on the status of collaborative experiments on basidiomycete tests. International Research Group on Wood Preservation, Document No. IRG/WP/2194. Stockholm, Sweden.

BRAVERY, A. F. & CAREY, J. K. 1977 Development of an improved technique for rapid screening of wood preservative fungicides: selection of test substrate. *International Biodeterioration Bulletin* **13**, (1), 18–24.

BRAVERY, A. F. & DERBYSHIRE, H. 1979 Control of blue stain fungi in service. *Materials and Organisms* **14**, (1), 55–80.

BRAVERY, A. F. & DICKINSON, D. J. 1985 Assessing the effectiveness of chemicals for preventing development of blue stain fungi in service: a co-operative study of artificial weathering systems. *Materials and Organisms* **20**, (2), 77–95.

BRAVERY, A. F. & MILLER, E. R. 1980 The rôle of pretreatments in the finishing of exterior softwood. *Records of the Annual Convention of the British Wood Preserving Association*, pp. 14–22.

BRITISH STANDARDS INSTITUTION 1968 Methods of test for fungal resistance of manufactured building materials. BS 1982:1968.

BRITISH STANDARDS INSTITUTION 1979 Accelerated ageing of treated wood prior to biological testing: evaporative ageing procedure (pr EN 73). BS 5761: Part 1: 1979.

BRITISH STANDARDS INSTITUTION 1980 Accelerated ageing of treated wood prior to biological testing: leaching procedure (pr EN 84). BS 5761: Part 2: 1980.

BRITISH STANDARDS INSTITUTION 1982 Wood preservatives. Determination of the toxic values against wood destroying Basidiomycetes cultured on an agar medium (pr EN 113). BS 6009: 1982.

BRITISH STANDARDS INSTITUTION 1986 Methods of test for fungal resistance of manufactured building materials made of or containing materials of organic origin. Part 1. Resistance to wood rotting Basidiomycetes. Draft for public comment. BS 1982 revised.

BUTCHER, J. A. 1973 Laboratory screening trials for new prophylactic chemicals against sap stain and decay in sawn timber. *Materials and Organisms* **8**, (1), 51–70.

BUTCHER, J. A. 1979 Testing new preservatives for protection of wood exposed in above ground situations. *Materials and Organisms* **14**, (1), 43–53.

BUTCHER, J. A. 1983 Facility for accelerated stake tests in unsterile soil. International Research Group on Wood Preservation Document No. IRG/WP/2169.

BUTCHER, J. A., PRESTON, A. F. & DRYSDALE, J. 1977a Initial screening trials of some quaternary ammonium compounds and amine salts as wood preservatives. *Forest Products Journal* **27**, (7), 19–22.

BUTCHER, J. A., HEDLEY, M. E. & DRYSDALE, J. 1977b Comparison of a quaternary ammonium compound and copper-chrome-arsenate as wood preservatives. *Forest Products Journal* **27**, (7), 22–25.

BUTIN, H. 1961 Ein neues Verfahren zur Bewertung der bläuewidrigen Wirksamkeit öliger Grundiermittel. *Holz als Roh und Werkstoff* **19**, 195.

CAREY, J. K. 1980 The mechanism of infection and decay of window joinery. PhD thesis, University of London p. 426.

CAREY, J. K. 1982 Assessing the performance of preservative treatments for window joinery. *Holz als Roh und Werkstoff* **40**, 269–274.

CAREY, J. K. & BRAVERY, A. F. 1985 Developments in the assessment of joinery preservatives. *Records of the Annual Convention of the British Wood Preserving Association*, pp. 3–11.

COMITÉ EUROPÉEN DE NORMALISATION 1984 Field test method for determining the protective effectiveness of a preservative in the marine environment. Doc. No. 38N 415E. May 1984.

COMITÉ EUROPÉEN DE NORMALISATION 1985a Test methods for wood preservatives. Laboratory method for determining the preventive effectiveness of a preservative treatment against blue stain in service, pr EN 152. July 1985.

COMITE EUROPÉEN DE NORMALISATION 1985b Field test method for determining the relative protective effectiveness of a wood preservative in ground contact. Doc. No. 38N 435E. April 1985.

COX, T. R. G. 1981 The influence of carrier solvents on the fungicidal efficacy of bis (tri-n-butyltin) oxide in wood. *International Journal of Wood Preservation* **1**, (4), 173–176.

CSERJESI, A. J. 1980 Procedure for field testing fungicides on unseasoned lumber. Technical Report No. 16, Forintek Canada Corp., January 1980.

CSERJESI, A. J. & JOHNSON, E. L. 1982 Mould and sap stain control: laboratory and field tests of 44 fungicidal formulations. *Forest Products Journal* **32**, (10), 59–68.

CSERJESI, A. J. & ROFF, J. W. 1970 Accelerated laboratory test for evaluating the toxicity of fungicides for lumber. *Materials Research and Standards* **10**, (3), 18–19, 59–60.

DEON, G. & TRONG, L. 1985 Resistance of board materials against fungal decay: a comparative experiment on Kollé flask test method and direct test method. International Research Group on Wood Preservation Document No. IRG/WP/2243.

DEPPE, H-J. & GERSONDE, N. 1977 Technological advances in the production and testing of preserved wood based panel products. *Journal of the Institute of Wood Science* **7**, (5), 20–25.

DICKER, P. E., DICKINSON, D. J., EDLUND, M. L. & HENNINGSSON, B. 1983 Borate diffusion techniques for the in situ treatment of joinery. *Records of the Annual Convention of the British Wood Preserving Association*, pp. 73–80.

DICKINSON, D. J. 1971 Disfigurement of decorative timbers by blue stain fungi. *Records of the Annual Convention of the British Wood Preserving Association*, pp. 151–169.

DICKINSON, D. J. 1974 A new technique for screening fungicides for wood preservation. *International Biodeterioration Bulletin* **10**, (2), 49–51.

DICKINSON, D. J. 1976 Proposals for a field experiment to determine the performance of preservative treated hardwoods with particular reference to soft rot. International Research Group on Wood Preservation Document No. IRG/WP/367.

DICKINSON, D. J. & HENNINGSSON, B. 1984 A field test with anti sap-stain chemicals on sawn pine timber stored and seasoned under different conditions. International Research Group on Wood Preservation Document No. IRG/WP/3245.

DICKINSON, D. J. & LEVY, J. F. 1982 Hardwood field experiment: Progress Report 1977–82. International Research Group on Wood Preservation Document No. IRG/WP/3200.

DREWS, G. 1976 *Mikrobiologisches Praktikum*. Berlin: Springer-Verlag.

DRYSDALE, J. A. 1984 Comparison of the effect of different soil sources on the type and rate of decay of CCA-treated pine exposed in a soil bed. International Research Group on Wood Preservation Document No. IRG/WP/2213.

DRYSDALE, J. A. & PRESTON, A. F. 1982 Laboratory screening trials with chemicals for the protection of green timber against fungi. *New Zealand Journal of Forest Science* **12**, (3), 457–466.

ESLYN, W. E. & CASSENS, D. L. 1983 Laboratory evaluation of selected fungicides for control of sap stain and moulds on southern pine timber. *Production Journal* **33**, (4), 65–68.

FINHOLT, R. W., WEEKS, N. & HATHAWAY, C. 1952 New theory on wood preservation. *Industrial and Engineering Chemistry* **44**, 101.

GERSONDE, N. & KERNER-GANG, W. 1976 A review of information available for development of a method for testing wood preservatives with soft rot fungi. *International Biodeterioration Bulletin* **12**, (1), 5–13.

GIBSON, E. J. 1966 The rôle of laboratory testing in the evaluation of wood preservatives. *Wood*, June 1966, **3**.

GREAVES, H. 1977 Potential toxicants for controlling soft rot in preservative treated hardwood. *Materials and Organisms* **12**, (4), 287–299.

GREAVES, H., MCCARTHY, R. & COOKSON, L. J. 1982 An accelerated field simulator trial of fused preservative rods. *International Journal of Wood Preservation* **2**, (2), 69–71.

HARTLEY, C. 1958 Evaluation of wood decay in experimental work. U.S. Forest Products Lab. Report No. 2119.

HATFIELD, I., SHUMARO, R. S., VAUGHAN, T. H. & HILL, E. F. 1950 The unseasoned wood disc method for evaluating sap stain control chemicals. *Phytopathology* **40**, 653.

HEDLEY, M. E. 1976 Preservative requirements for exterior particleboard predicted from accelerated laboratory tests. *New Zealand Journal of Forest Science*, **6**, (3), 455–460.

HEDLEY, M. E. 1982 Practical considerations in assessing the importance of biodeterioration of board materials and its prevention. *Biodeterioration 5*, eds Oxley, T. A. & Barry, S., pp. 117–126. London: John Wiley & Sons Ltd.

HEDLEY, M. E. 1983 Comparison of decay rates of preservative treated stakes in field and fungus cellar tests. International Research Group on Wood Preservation Document No. IRG/WP/2200.

HEDLEY, M. E. & FOSTER, J. B. 1972 Modified soil block technique for assessing wood decay. *New Zealand Journal of Forest Science*, **2**, (2), 237–248.

HEDLEY, M. E., PRESTON, A. F., CROSS, D. J. & BUTCHER, J. A. 1979 Screening of selected agricultural and industrial chemicals as wood preservatives. *International Biodeterioration Bulletin* **15**, (1), 9–18.

HEDLEY, M. E., TSUNODA, K. & NISHIMOTO, K. (1982) Evaluation of alkyl ammonium compounds for use as low toxicity wood preservatives in Japan. *Wood Research* **68**, 37–40. Kyoto University.

HELSING, G. G., MORRELL, J. & GRAHAM, R. D. 1984 Evaluation of fumigants for control of internal decay in pressure treated Douglas fir poles and piles. *Holzforschung* **38**, 277–280.

HENSHAW, B. G. 1978 In situ treatment of window joinery. *Timber Trades Journal*, 29 July 1978, pp. 19–20.

HIGHLEY, T. L. 1980 In place treatments for control of decay in waterfront structures. *Forest Products Journal* **30**, (9), 49–51.

HIGHLEY, T. L. 1984 In place treatments with water borne preservatives for control of decay in hardwoods and softwoods above ground. *Materials and Organisms* **19**, (2), 95–104.

HIGHLEY, T. L. & ESLYN, W. E. 1982 Using fumigants to control decay in waterfront structures. *Forest Products Journal* **32**, (2), 32–34.

HULME, M. A. & THOMAS, J. F. 1979 Control of fungal sap stain with alkaline solutions of quaternary ammonium compounds and with tertiary amine salts. *Forest Products Journal* **29**, (11), 26–29.

IMSGAARD, F., JENSEN, B., PLUMB, H. & LANDSIEDEL, H. 1985 Stability and performance of some modern fungicides used for wood preservation. *Records of the Annual Convention of the British Wood Preserving Association*, pp. 47–52.

INGLEBY 1982 New techniques for the evaluation of remedial treatment formulations in the laboratory. *Records of the Annual Convention of the British Wood Association*, pp. 39–43.

JACKSON, D. & SAVORY, J. G. 1968 The decay resistance of wood fibre building boards and particleboards. *International Biodeterioration Bulletin* **4**, (2), 83–88.

JOHNSON, G. C., THORNTON, J. D. & GREAVES, H. 1982 The accelerated field simulator (fungal cellar). International Research Group on Wood Preservation Document No. IRG/WP/2170.

KERNER-GANG, W. 1974 A contribution to the technique of testing preservatives against soft rot fungi by the vermiculite burial method. *Materials and Organisms* **9**, (4), 269–282.

KERNER-GANG, W. 1975 The use of the vermiculite burial method for testing preservatives against soft rot fungi. *Materials and Organisms* **10**, (1), 15–30.

LANCE, O. C. 1958 History and development of wood preservation for mill work. *Forest Products Journal* **8**, (10), 61A–65A.

LEA, R. G. 1980 In situ preservative treatments for joinery in service. British Wood Preserving Association News Sheet No. 162.

LEA, R. G. 1982 Contribution to the testing of wood based board materials. International Research Group on Wood Preservation Document No. IRG/WP/2176.

LEUTRITZ, J. 1946 A wood soil contact culture technique for laboratory study of wood destroying fungi, wood decay and wood preservation. *Bell System Technical Journal* **25**, 102.

LEVY, J. F. 1967 Decay and degrade of wood by soft rot fungi and other organisms. *Records of the Annual Convention of the British Wood Preserving Association*, pp. 147–160.

LEWIS, D. A., WILLIAMS, G. R. & EATON, R. A. 1985 The development of prophylactic chemicals for the treatment of green timber. *Records of the Annual Convention of the British Wood Preserving Association*, pp. 14–26.

MATEUS, T. J. E. 1954 Evaluation of the effectiveness of wood preservatives by a new method based on the measurement of deflections. Portugal Ministry of Public Works. National Laboratory Civil Engineers Publication, No. 48.

PURSLOW, D. F. 1979 Fifty years of field tests on wood preservatives at Princes Risborough. *International Journal of Wood Preservation* **1**, (1), 5–10.

PURSLOW, D. F. & WILLIAMS, N. A. 1978 Field trials on preserved timber out of ground contact. BRE Current Paper CP 78/78, Building Research Establishment, Princes Risborough Laboratory.

RABANUS, A. 1931 Über die Säure-Produktion von Pilzen und deren Einfluss auf die Wirkung von Holzschutzmitteln. *Mitteilungen der Fachauss für Holzfragen bei Deutsches Ingen und Deutsches Forstverein* **23**, 77.

RICHARDSON, B. A. 1978 Simplified Basidiomycete tests for wood preservative development work. International Research Group on Wood Preservation Document No. IRG/WP/2113.

SAUNDERS, L. D. A. & BLOW, D. P. 1985 Appraisal of fungicidal emulsions for in situ treatment. *Records of the Annual Convention of the British Wood Preserving Association*, pp. 64–72.

SAVORY, J. G. 1954a Breakdown of timber by ascomycetes and fungi imperfecti. *Annals of Applied Biology* **41**, 336.

SAVORY, J. G. 1954b Damage to wood caused by micro organisms. *Journal of Applied Bacteriology* **17**, 213.

SAVORY, J. G. 1969 The fungus resistance of standard hardboard. Forest Products Research Laboratory Timberlab Paper No. 15.

SAVORY, J. G. & BRAVERY, A. F. 1970a Collaborative experiments in testing the toxicity of wood preservatives to soft rot fungi. *Materials and Organisms* **5**, (1), 59–80.

SAVORY, J. G. & BRAVERY, A. F. 1970b Observations on methods of determining the effectiveness of wood preservatives against soft rot fungi. *Mitteilungen der Deutschen Gesellschaft für Holzforschung* **57**, 12–17.

SAVORY, J. G. & CAREY, J. K. 1976 Laboratory assessment of the toxic efficacy of joinery preservative treatments. *Records of the Annual Convention of the British Wood Preserving Association*, pp. 3–21.

SAVORY, J. G. & CAREY, J. K. 1979 Decay in external framed joinery in the United Kingdom. *Journal of the Institute of Wood Science* **8**, (4), 176–180.

SAVORY, J. G., CAREY, J. K. & STRIBBLING, M. A. 1977 Laboratory examination of simulated window joinery after outdoor exposure. *Materials and Organisms* **12**, (4), 315–326.

SCHEFFER, T. C. & GOLLOB, L. 1978 A bioassay for appraising preservative protection of wood above ground. *Holzforschung* **32**, (5), 157–161.

SCHEFFER, T. C. & GRAHAM, R. D. 1973 Bioassay estimation of sapwood protection in spray-treated western red cedar poles. *Forest Products Journal* **23**, (9), 106–109.

SCHEFFER, T. C. & LEW, J. D. 1976 Bioassay of residual preservative protection in wood. *Forest Products Journal* **26**, (7), 45–50.

SCHULZE, G. 1951 Ein mykologisches Verfahren zur Bewertung vorbeugeder Schitzmittel gegen das Verbläen von Kiefernholz. *Angewandte Botanik* **26**, 1.

SCHULZE, G. 1952 Ein mykologisches Verfahren zur Bewertung fungicider Grundiermittel mit bläuewidriger Wirkung. *Holz als Roh und Werkstoff* **10**, 353.

SMITH, D. N. R. 1965 The evaluation of wood preservatives. *Records of the Annual Convention of the British Wood Preserving Association*, p. 123.

SMITH, D. N. R. 1971 A possible method for the rapid evaluation of wood preservatives. *Mitteilungen der Deutschen Gesellschaft Holzforschung* **57**, 18–22.

SMITH, R. S. 1967 Carbon dioxide evolution as a measure of attack of wood by fungi and its application to testing wood preservatives and sapstain preventives. *Annals of Applied Biology* **59**, 473.

SMITH, R. S. & SAVORY, J. G. 1965 The toxicity of aged creosote to *Lentinus lepideus*. *Journal of the Institute of Wood Science* **13**, 31.

SUTTER, H-P. 1978 A new technique for screening fungicides for wood preservatives. International Research Group on Wood Preservation Document No. IRG/WP/2113, 71–80.

SUTTER, H-P. & CAREY, J. K. 1985 Bioassay to determine detoxification of tri-n-butyltin oxide by species of *Phialophora*. *International Biodeterioration* **13** (3), 195–200.

TAYLOR, J. M. 1978 Petri-dish plate contact test. (Personal communication.)

THORNTON, J. D. 1977 Potential toxicants for controlling soft rot in preservative treated hardwoods. *Materials and Organisms* **12**, (3), 201–210.

THORNTON, J. D. 1979 Evaluation of a new laboratory decay technique using *Serpula lacrymans*. *International Biodeterioration Bulletin* **15**, (2), 45–48.

THORNTON, J. D. & COLLETT, O. 1979 A laboratory test to determine potential uses of fungicides against *Serpula lacrymans* Grey. *International Journal of Wood Preservation* **1** (1), 21–25.

TRONG, L. 1984 A direct method for testing plywood and particleboards against fungal decay. International Research Group on Wood Preservation Document No. IRG/WP/2214.

VERRALL, A. F. 1961 Brush, dip and soak treatments with water repellent preservatives. *Forest Products Journal* **11**, (1), 23–26.

VINDEN, P., LEVY, J. F. & DICKINSON, D. J. 1983 Soil bed studies: the efficacy of wood preservatives. International Research Group on Wood Preservation Document No. IRG/WP/2205.

VINDEN, P., SAVORY, J. G., DICKINSON, D. J. & LEVY, J. F. 1982 Soil bed studies. International Research Group on Wood Preservation Document No. IRG/WP/2181.

WÄLCHLI, O. 1969 Die Prüfung der Wiederstandsfähigkeit von Holzschutzmitteln gegen Moderfäule. *Schweizer Archiv für Angewandte Wissenschaft Technologie* **35**, 73.

Mould Growth Testing of Materials, Components and Equipment to National and International Standards

JOAN KELLEY AND D. ALLSOPP
CAB International Mycological Institute, Ferry Lane, Kew, Surrey TW9 3AF

This paper is concerned with the testing of materials to recognized standards. Precise instructions and methods for carrying out tests have not been included as, by definition, these details appear in the published standards. It is intended to provide a critical review of current methods and to help guide those needing to use such techniques towards the available standards.

Modern equipments (electronic, optical, photographic, etc.) tend to have a high percentage of plastics materials either structural or as components, e.g. of printed circuit boards. Other materials (textiles, timbers, etc.) are discussed elsewhere in this publication, therefore this paper will cover mainly those standards which are designed to test plastics materials or completed pieces of equipment.

The production of materials to defined specifications and their testing to ensure standards are met, dates back to the last century. However, concern over the resistance of materials and equipment to attack by microorganisms only gained momentum during World War II when equipment designed in temperate zones was attacked on transfer to tropical areas. Since that period, knowledge of storage conditions and the use and variety of biocides has increased greatly. It is now relatively easy to predict and prevent, or at least delay, gross decay in many materials. Unfortunately this enhanced ability to retard deterioration has been paralleled by an increase in sophistication and miniaturization of, for example, electronic equipment and components, so that often the presence of even limited amounts of mycelium can still cause severe problems.

The prevention of microbial growth remains important today even with the advent of recalcitrant man-made molecules and the availability of biocides with enhanced activities. The problems have become more sophisticated, but many test methods remain little different from those designed in the early post-war years.

The Rationale of Testing Procedures

Jones (1968) summarized the aim of all microbial growth testing as 'the inducement

Industrial
Microbiological Testing
0-632-01793-7

of growth and measure of its degree'. To achieve this aim, the first requirement is to bring the material under investigation into contact with the micro-organisms which may attack it. This is usually brought about via pure culture inoculation, mixed culture inoculation or soil burial. Assessment of growth may be either direct (by observing the behaviour of the organism) or indirect (by determining physical or chemical changes in the substrate).

Depending on the information required, tests may run for different times, e.g. for 24 h to assess biocidal activities; for 28 days to determine visual changes and/or strength and weight losses; for 84 days for assessment of weight loss and other physical changes; for longer periods in the case of burial or exposure tests in which all forms of assessment may be involved.

Although it may be useful in general terms to test individual components of a plastics material, the possibilities of interactions make it important that the final product is also tested. For example, a susceptible plasticizer which does not migrate may be 'locked' within a non-degrading polymer rendering the final product resistant.

It is also important that a material's susceptibility to mould growth be considered early in the design stage. It is frequently assumed that, for example, plastics materials or glasses are all resistant to attack by micro-organisms. Problems can therefore occur at late stages in the design process after a product has passed through environmental testing. The discovery of unacceptable amounts of growth at this stage can cause delay and embarrassment.

Methods Employed for the Testing of Plastics Materials and Equipment

Pantke (1978) and Osman & Klausmeier (1977) have produced helpful reviews of the methods of testing plastics materials and plasticizers. Some of the methods generally employed, together with representative standards tests, are summarized below.

The use of solid agar medium

Examples of such usage are detailed in ISO 846 'Plastics: Determination of behaviours under the action of fungi and bacteria', and ASTM G-21-70 'Standard practice for determining the resistance of synthetic polymeric materials to fungi'.

The samples are placed on plates of medium which may or may not include a carbon source. Pre-weighed samples are either first dipped into spore suspensions or they may be placed on pre-grown lawns. The susceptibility of the material to attack or the presence of any biostatic/biocidal capability can then be assessed. Visual assessment is carried out along with weight loss determinations.

Respirometry has been employed with solid media (Burgess & Derby 1964).

The sample and organisms are placed on the medium in the flask of a Warburg manometer and oxygen uptake measured.

The use of liquid medium

An example of the use of a liquid medium appears in NFX 41-513, 'Resistance of plasticizers to attack by micro-organisms'.

This method is used to evaluate the resistance of plasticizers to fungal attack and to determine any fungistatic effects of the formulation. The material is added to a mineral salt solution and the fungal growth compared to that in a castor oil control.

Growth assessment is visual only. Other methods of assessing growth in liquid tests are: cell dry weights (Pankhurst *et al.* 1972), viable cell counts (by serial dilution or most probable number techniques); turbidity measurements, respirometry and determination of endogenous metabolism (Sharp & Woodrow 1972).

Soil burial

An example of the use of soil burial can be found in DIN 53 739, 'Determination of the resistance of plastics to fungi and bacteria.'

Soil burial methods are severe and are usually employed on materials that come into contact with the ground, for example construction materials and tentage. The method appears in a number of standards. The test material is buried under laboratory or field conditions for a specified time. Exhumed materials are subjected to visual examination and to assessment of changes in weight and tensile strength.

Respirometry has also been employed to assess the condition of exhumed samples (Waite 1978).

Environmental chamber tests

Examples of this type of test can be found in BS 2011 p. 2.1 J 'Basic environmental testing procedures—Mould growth' and Mil Std 810D p. 508.3.

The samples are sprayed or dipped in a mixed spore suspension and incubated with or without an added nutrient source at raised temperatures and high relative humidities for 28 or 84 days. Samples are usually assessed visually but operational tests and the measurement of physical parameters may also be carried out. These tests are often used to assess the resistance of pieces of equipment and completed items.

Tropical chamber test

An example of this type of test is employed in NFX 41-515 'Methods of testing the resistance of plastics materials to attack.'

Samples are placed in a tropical test room at raised temperatures and relative humidities (these are often cycled). Chambers may either be inoculated regularly with spores or trays of soil may be left in the chambers as an inoculant. This method is often used in long-term storage trials. Degradation can be assessed by most of the methods previously discussed.

Maintenance and Monitoring of the Environment

Relative humidity and temperature

A constant, defined relative humidity (RH) is usually required in standards based on chamber or tropical storage tests, and can be brought about by the following methods. (1) The use of a purpose-designed environmental test chamber or room with control based on salt solutions or steam generation to create the appropriate RH. (2) The use of a normal incubator with the specimens held within a closed container supported over a salt solution or water depending on the RH limits specified.

On a routine basis RH can be measured using a hair hygrometer, although these must be calibrated before use and checked regularly during the test. Single readings can be taken using electronic probes but continual exposure of these to high humidities often results in saturation of some types.

Temperatures can be measured using an RH/temperature probe, a thermocouple or, most simply, by placing a thermometer in the test tank. It should be remembered that if operating to National Testing Laboratory Accreditation Scheme (NATLAS) standards all monitoring equipment must be calibrated and traceable to national standards.

A typical 'basic' set up is illustrated in Fig. 1. The plastic tank has water in the bottom and samples are placed on stands. A thermometer and hygrometer are suspended within the tank which is then covered. Domestic food wrapping film has been found to be useful for this purpose. The whole is then placed in a temperature-controlled incubator for the specified period.

Air flow

The mould growth test in Mil. Std 810 D part 508.3 (1983) sets problems in that it requires a defined flow of air over the sample and also the temperature to be cycled. Environmental test chambers often include a fan in the design to circulate the air. However, smaller laboratories using traditional biological incubators may find this requirement difficult to fulful. Fig. 2 illustrates a tank employed by the Biodeterioration Laboratory, Directorate of Quality Assurance/Technical Support (DQA/TS), Royal Arsenal East, London SE18 6TD. The tank has a small fan which provides the correct airflow and the whole may be placed within an incubator, or, alternatively, a small heater can be placed within the tank to maintain the temperature.

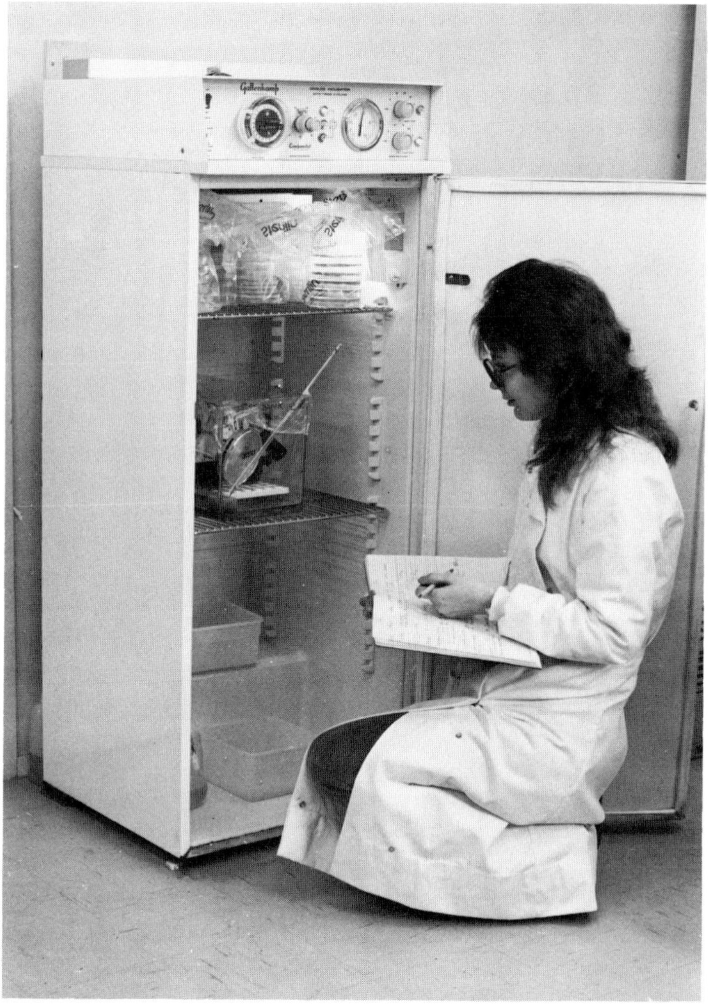

FIG. 1. A 'basic' set-up for testing to BS 2011 p. 2.1 J.

Test organisms

The fungi employed in the testing of materials have in some cases been isolated from the type of material they are designed to test. However, many of them have been adopted from other fields (for example, textile testing organisms used for testing plastics materials); the reason behind the original selection of many organisms is now uncertain.

Table 1 shows the organisms employed in a number of standards. It illustrates that many isolates are common to several tests. It is unlikely that these organisms

FIG. 2. A tank employed for mould growth testing to Mil. Std. 810 D part 508.3.

were selected for each test to give the optimum 'performance' and the choice was probably made by selecting from other earlier tests. Furthermore, several of the isolates used in plastics testing for example were not isolated from man-made polymeric materials (Table 2). Those that were obtained from plastics were also probably never tested to ensure they possessed the enzyme activities necessary to actually break down those materials. A number of these organisms were first isolated 40–50 years ago, and were not deposited in a public culture collection until some years after isolation. Organisms may have changed their characteristics during 50

TABLE 1. *Some organisms used in standard testing*

Standard	*Aspergillus niger* 17454[1]	*Aspergillus niger* 45551	*Aspergillus niger* 91855	*Aspergillus terreus* 45543	*Aspergillus amstelodami* 17455	*Aureobasidium pullulans* 45533	*Aureobasidium pullulans* 145194
BS 2011 p2.1j	✓			✓		✓	
160–500–15	✓				✓		
ASTM G21–70			✓				
BLS 62.21.b	✓				✓		
BS 1982 : p.3							
BS 2848	✓				✓		
BS 3046	✓						
BS 4249	✓				✓		
BS 5980	✓						
BS 6085		✓		✓			
ISO 846		✓					✓
D 3273		✓					
Def 07–55	✓					✓	✓
Mil 810 D			✓				
TT-P-29H		✓					✓

Standard	*Paecilomyces variotii* 108007	*Penicillium funiculosum* 114933	*Penicillium ochrochloron* 6127li	*Scopulariopsis brevicaulis* 49528	*Trichoderma viride* 45553i	*Chaetomium globosum* 45550	*Chaetomium globosum* 16203
BS 2011 p2.1j	✓	✓	✓	✓	✓		
160–500–15	✓						✓
ASTM G21-70					✓	✓	
BLS 62.21.b	✓						✓
BS 1982 : p.3	✓	✓					✓
BS 2848	✓						✓
BS 3046							✓
BS 4249	✓						✓
BS 5980							
BS 6085	✓	✓	✓	✓	✓	✓	
ISO 846		✓			✓	✓	
D 3273							
Def 07–55	✓	✓	✓	✓	✓		
Mil 810 D						✓	
TT-P-29H							

[1] IMI numbers

TABLE 2. *Source of some organisms used in standards testing*

Used in standard designed for plastics testing	IMI No.	Name given in test	Correct identification	Original substrates	Year of isolation
No	17454	*Aspergillus niger*	*A. niger*	?	1947
Yes	45551	*Aspergillus niger*	*A. niger*	Leather	1951
Yes	91855	*Apsergillus niger*	*A. niger*	Radio set	1944
No	45543	*Aspergillus terreus*	*A. terreus*	Haversack	1951
No	17455	*Aspergillus amstelodami*	*A. amstelodami*	?	1947
No	45533	*Aureobasidium pullulans*	*Aur. pullulans*	Painted wood	1951
Yes	145194	*Aureobasidium pullulans*	*Aur. pullulans*	Plastic	1969
Yes (1)	108007	*Paecilomyces variotii*	*Paecilomyces variotii*	Synthetic rubber	1964
Yes	114933	*Penicillium funiculosum*	*P. pinophilum*	P.V.C.	1960
Yes (1)	61271	*Penicillium ochrochloron*	*P. ochrochloron*	?	1955
Yes (1)	49528	*Scopulariopsis brevicaulis*	*S. brevicaulis*	Chrysalis of silk worm	1952
Yes	45553	*Trichoderma viride*	*Gliocladium virens*	Soil but grown on plastics	1944
Yes	45550	*Chaetomium globosum*	*C. globosum*	Cotton	1933
No	16203	*Chaetomium globosum*	*C. globosum*	Paper	1942

(1) = Used for cable sleeve tests—these are mainly plastics materials.

years of maintenance. Much of this time may have elapsed before freeze drying and liquid nitrogen storage were adopted.

Variation in the genome may be brought about by spontaneous mutation, heterokaryosis and the parasexual cycle or selection during transfer. Isolates can therefore lose their ability to degrade the test substrate. It is essential that once selected, test organisms should either be freshly recovered from freeze dried or liquid nitrogen stored material at regular intervals or checked for relevant enzyme activity regularly. CAB International Mycological Institute has now established a biochemical facility so that it can confirm the activities of these cultures prior to use in tests.

A number of organisms which are employed in testing have been wrongly named. For example, IMI 45553, originally named and listed as *Trichoderma viride*, is *Gliocladium virens*, and IMI 114933 is not *Penicillium funiculosum* but *P. pinophilum*.

As standards are re-written and updated, authorities should be re-examining the isolates cited, their enzyme activities, and the methods of test in general. For example, the new IEC 68-2-10 (International Electrotechnical Commission) (BS2011 p. 2.1 J) was produced following round robin trials in five European laboratories on 'standard' circuit boards. Also, the International Biodeterioration Research Group (IBRG) is 'performance testing' the new DIN 53 739 Standard, revitalising work carried out by IBRG in its early years on plasticized PVC. This international approach to standard design has gained ground over the last two decades and is to be encouraged. However, closer scrutiny of the organisms employed in tests is also desirable because many are still based on isolates first selected 40 years ago.

Factors Affecting Testing

Characteristics of the Inoculum

Single or mixed inoculum

Hazeu (1967) and Klausmeier (1972) reported on interlaboratory studies comparing the use of mixed inocula as opposed to pure cultures in agar plate testing for the microbial susceptibility of plastics. Hazeu (1967) compared three common fungal test mixes and found that they produced essentially identical changes in plastics formulations. Klausmeier (1972) found that three organisms included in each of these mixes, *Aspergillus niger*, *Penicillium funiculosum* and a *Trichoderma* species, all caused weight losses identical to those brought about by mixed culture. In these cases there does not appear to be any inhibition or antagonism between the fungi, a possibility which must be considered when selecting test organisms. For example, some *Chaetomium* and *Gliocladium* species have strong antifungal properties (Domsch *et al.* 1980).

When biocide treated materials were tested in the same manner, single inocula showed preferential deterioration of formulations depending on the efficacy of the biocide. Klausmeier (1972) concluded therefore that it was not advisable to use pure culture techniques when ingredients having possible biocidal activities were included in the formulation. Different organisms have different susceptibilities to biocides and single culture tests can therefore provide misleading results.

Size of inoculum

Many plate and environmental chamber test methods require a minimum number of spores/ml of test organism suspension. Jones (1968) felt that inoculum size had a marked effect on growth, quoting Meyrath (1962) who found growth rate and mycelium yield to be larger with larger inocula. However, providing sufficient minimum inoculum is present, the quantity is probably not critical. Osman & Klausmeier (1971) showed that concentrations above 10^4 spores/ml produced no differences in the degradation of plastics.

Washing the inoculum

Most test methods where the substrate is to be the only carbon source now require spores to be washed before use. This prevents the carry over of nutrients from the growth medium and from any cells which may be lysed during the preparation of the spore suspensions. It also removes any fungistatic material present in the spore suspensions.

Arrangement of Samples

This mainly applies to solid media testing where the samples are arranged on agar in a petri dish. Klausmeier (1972) compared weight losses in plastics samples incubated singly, as multiples of one sample type and as mixed multiple sample types. Deterioration was most apparent on those incubated singly, and those where all samples were of the same composition showed the lowest deterioration.

Although these samples had been treated with biocides, similar results have been obtained with untreated polyurethane samples by the International Biodeterioration Research Group (IBRG). Sample arrangement is therefore of importance and must also be taken into account when the material is to be subjected to tensile strength testing after either exposure on solid medium or after soil burial. It is important that stresses are not set up in the materials by improper placement or bending.

Shape and Size of Sample

Osman *et al.* (1972) showed that the shape and size of a sample can influence the results of biodeterioration testing significantly.

(1) A sample placed on the surface of a solid medium prevents the diffusion of oxygen to the area covered by the sample. Growth of the test organism may therefore be confined to the perimeter of the sample.

(2) Plasticizers and similar materials will diffuse through the material to the edge and hence the greater the perimeter, the greater the area of the sample exposed to attack.

When the perimeter to surface area was maximized the rate of biodeterioration was found to be maximized.

These workers demonstrated that degradation occurs from the perimeter of the sample inwards. They sliced a sample of PVC that had undergone plate testing and measured the modulus of elasticity of each strip. The inner strips retained their elasticity for a much longer period than the outer ones. These results, together with evidence for oxygen shortage, prompted them to suggest that: (1) samples for weight loss testing should be no larger than $1 \times 100 \times 1$ mm; (2) if the sample size must be larger, incubation should be in shake flasks to maximize the interface at which plastic, nutrient salts, and oxygen are available.

They also employed a technique in which the sample was inoculated with a seeded slice of medium on top of the material. This method removed the occurrence of 'edge' effect.

Problems Facing the Assessor

The Choice of the Method of Assessment

Most current visual assessment regimes are subjective and definitions of percentage cover are open to different interpretations.

Jones (1968) suggested that graded photographs relevant to the test materials could be reproduced in standards to act as comparisons. However, the problem of obtaining representative photographs of the great variety of growth forms which would be valid for the range of materials investigated, would be immense.

Tests requiring visual assessment alone usually run for 28 days; surface damage can then be investigated by light and scanning electron microscopy. Assessments are recorded as percentage coverage on the basis of scales which are published in the standards.

Many plastics and timber preservative standard tests have a weight loss assessment after plate testing or soil burial. The main problem here is removing microbial growth from solid substrates before weighing, but this is still recognized as the most accurate assessment of many materials' susceptibility to attack.

Long-term effects, determined either by plate or soil burial tests, are often assessed by tensile strength testing. Problems here are associated with the physical technique, those of conditioning the samples, jaw type and size, sample extension, etc. Multiple samples are necessary for statistical validity, as, unless gross damage has occurred, it is often difficult to detect significant differences between test materials and sterile controls.

Respirometry can produce rapid results, but, unless conditioned organisms are employed, misleading results can be obtained. The effect of constitutive enzymes only will be assessed and those of induced enzymes missed entirely. Consequently this is not a method which should be used in isolation.

Post-test changes can also be assessed by extraction followed by thin layer chromatography or gas liquid chromatography. Infra-red spectroscopy, measurements of degree of polymerization, and changes in glass transition points are examples of other useful techniques for investigating the attack of plastics materials. These methods, like weight loss, produce objective results, but interpretation is necessary as to the real effect on the material.

Selection of Controls

Most test controls are in effect a viability test and a check that the environmental conditions permit growth. The use of fully comparable controls when physical parameters are to be measured is extremely difficult. When attempting to check the functioning of equipment after testing, adequate controls become imperative yet are at the same time almost impossible to provide. If, for example, a piece of electrical equipment is to be function-checked before and after testing, a control exposed to the same raised temperatures and relative humidity should be provided. This control must therefore be sterilized and remain sterile throughout the test. Often surface sterilization with 70% alcohol or Hg_2Cl_2 is not possible because of the damage it would cause to the substrate, or insufficient because of the inaccessibility of internal components. Autoclaving is not impossible, but fumigation or in some cases irradiation are the best available methods.

Assuming sterilization is possible, the equipment must also be maintained in a sterile condition. This requires a second incubation chamber which must be pre-sterilized and difficulties can arise in maintaining sterility.

A greater problem arises in carrying out invasive tests where nutrients are sprayed on to equipment. It is vital that the control is sprayed with the nutrient solution in case corrosion or other adverse effects are enhanced. However, adding the presence of nutrients to the problems of sterilization mentioned above makes it virtually impossible to keep the control free from growth. All a tester can do is use his judgment when assessing the controls and trust that the sterilization attempts have kept microbial development to an insignificant level. With plastics materials, the need for controls is great, as the conditions which enhance mould growth can also result in chemical attack, causing oxidation and/or hydrolysis of the materials. Nylon materials are particularly susceptible to hydrolysis under test conditions so that adequate controls are essential.

Lack of Information

The work of an assessor is often made more difficult by a lack of information in three areas:

(1) It is often not possible to find out the composition of a plastics material received for testing. If the nature of the basic polymer is revealed, and if it is known to be resistant, then weight losses can be expressed as a percentage loss of plasticizer. plasticizer. The manufacturer will then be in a position to say what effect that percentage loss of plasticizer would have on the material. If a rough formulation is

known, then, even without attempts at quantification, it is possible to suggest causative agents when biodeterioration is observed.

(2) Clear information is rarely given on the final use of a material and on what information the manufacturer hopes to get from the test. Often a material which is moderately susceptible to mould growth will pose no problem if it is to operate, for example, in a closed environment under an inert gas atmosphere. However, that same material should not be considered as, say, a sealant on aircraft hydraulic systems expected to operate in the tropics.

(3) In the case of environmental chamber testing of equipment, mould growth forms part of a complete programme of tests including imposition of, for example, hot dry, hot wet, salt spray, environmental conditions. Although a single piece of equipment is often dedicated to the mould growth test, sometimes it is not, and there is little hope of obtaining the history of the equipment, even though previous treatment, for example, accelerated weathering tests, may greatly affect the results of the mould growth trials.

Reappraisal of Test Methods

Most of the tests described here originate in long-established standards. Updates occur, but usually result in modification rather than a complete reappraisal of the methodology. This may be the only practical way in which to proceed, but it is disturbing to note that although much of the experimental work on plastics materials described here was carried out by Klausmeier and the IBRG in the early 1970s, the results of this work have still not been incorporated into most national standards.

The measurements of change in physical parameters following microbiological attack are often reliable methods of assessment. However, visual assessment is subjective and therefore unreliable in the hands of untrained personnel. There is clearly a need for new methods of quantifying growth on surfaces. Methods employing fluorescent stains and light emission from the luciferin-luciferase reaction are under investigation at various centres but all have disadvantages at the moment (McCarthy 1984).

One of the difficulties encountered when trying to gather information together on microbiological standards testing is actually finding the test methods. Many (particularly environmental test chamber types) are included in large volumes of general environmental test methods, as in BS 2011.

CAB International Mycological Institute is collecting information on standards which include tests of material involving micro-organisms. At the moment this is a small data base held on a microcomputer. It is hoped, however, to expand it to a fully searchable facility on our VAX 11/780.

References

American Society for Testing Materials 1980 Standard practice for determining the resistance of synthetic polymeric materials to fungi. ASTM G-21-70.

Association Française de Normalisation 1961 Resistance of plasticizers to attack by microorganisms NFX 41-513.

Association Française de Normalisation 1962 Methods of testing the resistance of plastics materials to attack NFX 41-515.

British Standards Institution 1985 Basic environmental testing procedures part 2.1. Test J. Mould growth. BS 2011 p.2.1. J.

Burgess, R. & Darby, A. E. 1964 Two tests for the assessment of microbial activity on plastics. *British Plastics* **37**, 32–37.

Deutsch Institut für Normung (Draft) Determination of the resistance of plastics to fungi and bacteria by visual examination and measurement of changes in mass and physical properties. DIN 53 739.

Domsch, K. H., Gams, W. & Anderson, T-H 1980 *Compendium of Soil Fungi*, Vols 1 and 2. London: Academic Press.

Hazeu, W. 1967 Results of the first interlaboratory test on plastics. *International Biodeterioration Bulletin* **3**, 15–19.

International Electrotechnical Commission 1985 Basic environmental testing procedures. Test J. Mould Growth. IEC 68-2-10.

International Standards Organization 1978 Plastics: determination of behaviours under the action of fungi and bacteria—Evaluation by visual examination. ISO 846.

Jones, E. S. L. 1968 Some problems posed by quality screening for biodeterioration. *Biodeterioration of Materials*, eds Walters, A. H. & Elphick, J. J. pp. 188–195. Elsevier Publishing Co.

Klausmeier, R. E. 1972 Results of the second interlaboratory experiment on the biodeterioration of plastics. *International Biodeterioration Bulletin* **8**, 3–7.

McCarthy, B. J. 1984 Bioluminescent determination of microbial activity on textiles. Third International Symposium on Analytical Applications of Bioluminescence and Chemiluminescence.

Meyrath, J. 1962 Size of inoculum and growth kinetics of moulds. *Experientia*, **18**, 41.

Osman, J. L. & Klausmeier, R. E. 1971 *Candida lipolytica*: a non pathogenic test organism for deterioration of plastics. *Developments in Industrial Microbiology* **12**, 197–203.

Osman, J. L. & Klausmeier, R. E. 1977 Techniques for assessing biodeterioration of plastics and plasticizers. *Biodeterioration Investigation Techniques*, ed. Walters, A. H. pp. 77–94. London: Applied Science Publishers.

Osman, J. L. Klausmeier, R. E. & Jamieson, E. I. 1972 Rate limiting factors in biodeterioration of plastics. *Biodeterioration of Materials*, eds Walters, A. H. & Hueck-van-der Plas, E. H. pp. 66–75. London: Applied Science Publishers.

Pankhurst, E. S., Davies, M. J. & Blake, H. M. 1972 The ability of polymers or materials containing polymers to provide a source of carbon for selected microorganisms. *Biodeterioration of Materials*, Vol. 2, eds Walters, A. H. & Hueck-van-der Plas, E. H. pp. 76–90. London: Applied Science Publishers.

Pantke, M. 1978 Test methods for evaluation of susceptibility of plasticised PVC and its components to microbial attack. *Biodeterioration Investigation Techniques*, ed. Walters, A. H. pp. 51–75. London: Applied Science Publishers.

Sharp, A. N. & Woodrow, M. N. 1972 A rapid test for biodegradability using *Pseudomonas* organisms.

Biodeterioration of Materials, Vol. 2, eds Walters, A. H. & Hueck-van-der Plas, E. H. pp. 233–237. London: Applied Science Publishers.

U.S. DEPARTMENT OF DEFENSE 1983. Mil Std. 810 D. 1983. Method 508.3. Fungus.

WAITE, J. 1978 Some aspects of the microbiological properties of polymers and composite materials. PhD. thesis, University of Aston in Birmingham.

Microbiological Testing of Textile Materials

B. J. McCarthy, P. S. Senior and P. G. H. Bakker*
Wira Technology Group, Wira House, West Park Ring Road, Leeds LS16 6QL, and
**International Wool Secretariat, Valley Drive, Ilkley LS29 8PB*

Natural and man-made fibre production and subsequent textile processing represents a major global industrial sector. For example, within the European Economic Community, the Textile Industry provides employment for 2.7 million people and produces an annual turnover of approximately £53 billion. In recent years, growing demands for industrial textiles have supplemented existing apparel and household textile markets and have lead to increasing requirements for hygienic and microbial deterioration testing. Various novel techniques have been introduced for bioburden assessment on textile materials and for monitoring the performance of textile-associated biocides. Such developments have concentrated on reducing testing duration and attempting the introduction of non-destructive evaluation (N.D.E.) procedures. As the various major national biodeterioration test standards have been reviewed elsewhere (McCarthy 1984), the present chapter will highlight techniques for assessing microbial contamination on textile materials, with special reference to methods for the non-destructive testing of bioburden on fabrics and carpets, and for monitoring the survival of selected challenge organisms and natural populations on test substrates for hygienic evaluations.

Methods

Two basic types of microbial enumeration methods have been described: (i) nondestructive, which leaves the substrate essentially unaltered after testing, and (ii) destructive, which alters the test substrate irrevocably. The following test procedures have been applied to textile fabric and/or carpet substrates for general bioburden assessments and, where appropriate, to time-dependant survival studies. Hospital isolates and culture collection strains (identified where appropriate in the present text) may be used as challenge organisms. Recovery and enumeration media will be determined by organism requirements.

Industrial
Microbiological Testing
0–632–01793–7

A. Non-destructive Evaluations

The requirement for qualitative and semi-quantitative information concerning microbial contaminant levels on carpets and fabrics has lead to the development of a wide range of experimental procedures, ranging from simple techniques requiring a minimum of equipment, to complex highly-specialized devices. These procedures have been applied to hygienic studies of hospital textiles and carpets and to general assessments of anti-microbial efficacy. The methods are limited to studies where surface contamination, or the effects of mechanical disturbance leading to the release of microbial aerosols, are of importance. Some of the methods described may be modified to suit available equipment.

Direct Surface Recovery Methods

(i) *Contact plating*
Rubbo & Dixson (1960) described a simple portable device consisting of an aluminium disc approximately 2 mm thick and 85 mm in diameter mounted on a wooden handle 90 mm long. The sample of fabric is stretched over the disc which is then firmly pressed onto the surface of a standard 90 mm agar dish. Single or triple contacts may be made with the same device. A triple contact involves taking samples from three separate locations on the fabric using the same agar dish and is recommended for the detection of microbes which may be present in low numbers (e.g. *Staphylococcus aureus*). Pilsworth (1969) outlined a related method using wooden discs covered with sterile velvet for obtaining impression contact samples of organisms on the surface of carpets. Prepare 15.2 cm square sections of test carpet and pin to mounting boards. Inoculate the centre of the specimen with 2.5 or 5.0 ml of a bacterial suspension (e.g. *S. aureus* or *Shigella sonnei*) and allow to soak in for 5 min. Remove excess fluid with clean cellulose tissue or a pasteur pipette. Press the disc/velvet pad onto the test carpet and then onto an appropriate agar plate. For survival trials, repeat the process at suitable time intervals. Pilsworth used the method to compare survival rates on biocide-treated and untreated carpets and felts. Contact plating and Rodac plates (Nicholes 1970; Wetzler 1971) will again only recover microbes from the surface of the material. The agar sausage technique (ten Cate 1965) and the surface swab technique (Colbeck 1956) face similar limitations. Suzuki *et al.* (1984) have examined bacterial contamination of floors and other surfaces in operating rooms using contact plates. Carpeted areas were examined using stamp agar impression plates (Eiken Kizai Co., Tokyo, Japan) combined with brain–heart infusion agar or Eiken Stamp media containing brain–heart infusion, Tween 80, Lubrol W, lecithin, and thiosulphate as residual disinfectant neutralizers.

(ii) *Sweep plating*
Sweep an inverted petri dish containing an appropriate culture medium over the surface of the fabric or carpet sample. The leading edge of the plate flattens the fibres as it passes over them; the organisms are released and collected on the inverted agar surface. The test is difficult to standardize and is further complicated by the congregation of colonies growing at the leading edge of the plate forming confluent areas making counting procedures difficult (Blowers & Wallace 1955; Rubbo & Dixson 1960).

(iii) *Liquid recovery*
Dispense 2 ml of sterile tap water onto the surface of the test specimen. Using a pasteur pipette, recover residual liquid and plate out on appropriate nutrient media. This method is crude and inaccurate and should only be used for demonstrating gross changes in microbial numbers during such processes as carpet vacuuming (Litsky 1973).

Percussion Sampling

Initially suggested by Puck *et al.* (1946) and modified by Rubbo & Dixson (1960), the apparatus consists of a metal cylinder standing on an enclosed hollow base. Place an appropriate agar plate in the base and stretch the test fabric tightly over the bottom of the cylinder. Insert the cylinder into the base so that the test fabric is approximately 50 mm above the agar surface. Using a 250 g steel ball, dislodge the microbial contaminants by dropping the ball down the hollow cylinder onto the fabric. The height and number of drops and the subsequent settlement time may be varied.

Combined Vacuum/Agitation Sampling

(i) *Folin aeration apparatus*
Connect a standard Folin aeration apparatus at one end of a 400 mm rubber tube to a 75–100 mm glass funnel at the other. Using a vacuum pump operating at a flow rate of 25–30 l/min, sweep the funnel gently over the surface of the test specimen for 2 min. Collect the incoming air in 20 ml of an appropriate broth containing two drops of sterile olive oil to prevent foaming. Plate out aliquots of the resultant broth to obtain colony counts (Lemon 1943).

(ii) *Sieve sampling*
This method requires a sieve sampler developed by the Technology Branch of the Communicable Disease Centre, Public Health Service and Biological Warfare Section, U.S. Army, Fort Derrick, U.S.A., which consists of an aluminium holder covered by a sieve. Place a petri dish containing suitable nutrients in the apparatus

and then invert onto the surface of the textile substrate. Connect the device to a vacuum pump operating at a rate of 2 ft^3 of air per minute, and gently move the sieve sampler over the required area to be sampled. Incubate the plates as required (Schaffer 1966).

(iii) *Probe sampling*
A modification of the above procedures involves air being drawn through a cannula by a vacuum pump prior to collection by a glass impinger whilst guided by a template. Prepare a plexiglass template measuring 43.6 × 40.6 cm. Cut thirteen slits into the template measuring 33 × 0.64 cm and 1.9 cm apart. Assemble the template and an all glass liquid impinger onto a holder and connect an aluminium cannula and vacuum pump. Walter & Stober (1968) outlined the following experimental procedure: place the template on the carpet or fabric surface and insert the cannula into the holder. Place 40 ml of sterile phosphate buffered distilled water (pH 7.0) in the impinger and connect the vacuum pump. Draw the sampling cannula over the surface of substrate at the required depth (e.g. pile height) following the slit guidelines. Sample all thirteen slits in approximately 60 s. Flush the device by passing 60 ml of sterile buffered distilled water through the cannula. Add 0.1 ml sterile Tween 80 to the resultant 100 ml sample and sonicate at 25 kHz for 3 min to break microbial clumps. Prepare serial dilutions and plate onto appropriate media. Express results as colony forming units per square centimetre.

(iv) *Portable beater sampling*
This device comprises a modified vegetable chopper attached to a slit sampler. Introduced by Litsky (1973), the dome of the apparatus contains chopping blades which are rotated by an external handle to mechanically beat the substrate. Run the slit sampler using appropriate recovery media for 5 min to determine background levels. Operate the beating mechanism and slit sampler simultaneously for 5 min to obtain the specimen reading. Take a further slit sample for 10 min after the beating has stopped.

(v) *Carpet beater sampler*
Devised by the International Wool Secretariat, this device is intended to simulate the effect of traffic over the carpet under test and to assess subsequent aerial dispersion of microbial contaminants. The sampler consists of a glass or plastic tank of about 7 litres capacity (240 × 170 × 170 mm) and a rectangular metal 'foot' with a smooth bottom measuring 175 × 150 mm and weighing 475 g (see Fig. 1). Place a carpet specimen (240 × 170 mm) inside the base of the tank or place the sampler onto a fitted carpet for *in situ* measurements.

Place two settle plates containing appropriate nutrients in the tank. To simulate walking, raise the foot to its maximum elevation (150 mm) and allow to drop. Repeat the process to the required number of replicates (ten footfalls is a figure that

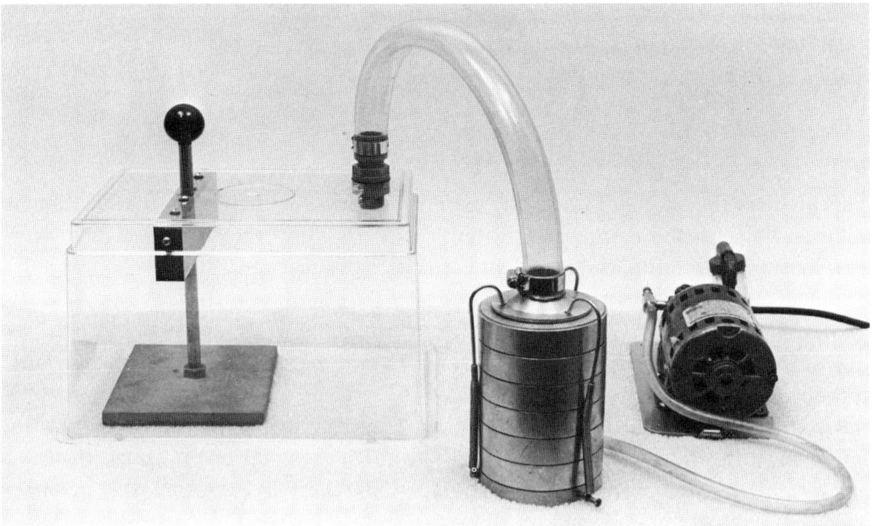

FIG. 1. General view of the apparatus showing the carpet beater sampler attached to an Andersen Atmospheric sampler.

has been used in various studies) and allow the air to settle for 10 min. Initial control results may be obtained by exposing comparable settle plates for 10 min with the foot stationary. Swab the tank and metal foot with 70% ethanol between sampling sites or samples. The following equation may be used to calculate the number of microbes per 10 cm^2:

$$\frac{\text{No. of bacteria per plate} \times \text{Surface area of carpet} \times 10}{\text{Surface area foot} \times \text{Surface area plate}} = \text{cfu}/10\ \text{cm}^2$$

Surface areas are expressed in square centimetres. The method has been applied to studies involving vacuum and shampoo carpet treatments and to the evaluation of the number of microbes on carpets treated with anti-microbial agents and their respective controls (Bakker, unpublished results).

A joint W.T.G./I.W.S. study is in progress to extend the range of application of the device by assessing contaminant dispersal from the substrate by means of an Andersen Multi-Stage Atmospheric Sampler (Atlanta, Georgia, U.S.A.) coupled directly to the top of the tank.

B. Destructive Evaluations

Quantitative studies generally require destructive sample collection and/or sample preparation. The following procedures may be used to quantify microbial population levels on carpets and fabrics and, where necessary, to monitor the survival of challenge organisms.

Maceration Sampling

This involves the homogenization of fabric or carpet specimens in a suitable diluent followed by conventional plating procedures. Variations on this method have been used to monitor microbial survival on both types of substrate.

(i) *Maceration—survival of microbes on fabrics* (Wilkoff *et al.* 1969a, b)
Cut seventy-five fabric swatches (approx. 5 cm diameter) and inoculate with appropriate suspensions of test organisms (e.g. *Staphylococcus aureus* or *Salmonella typhimurium*). Inoculation may be by direct contact, infected dust, or by aerosol spraying. Store the swatches in units of five in elevated lid petri dishes, ensuring separation by using sterile glass-fibre partitions. Test five specimens (one unit) immediately to obtain a Time 0 reading. Incubate the remaining petri dish units at 25°C and relative humidity of 35% (or alternatively 78%). Monitor residual bacterial populations at weekly intervals. Test each fabric specimen by cutting into small pieces and adding to 25 ml physiological saline. Homogenise for 30 s (e.g. using a Sorvall Omnimix Homogenizer, Ivan Sorvall Inc., Norwalk, Conn., U.S.A.). Prepare tenfold dilutions from the original suspension. Plate 0.2 ml aliquots onto appropriate media.

(ii) *Maceration—survival of microbes on carpets* (Walter & Stober 1967)
Walter & Stober (1967) exposed a variety of 45.7 cm^2 test carpets in a busy thoroughfare for periods ranging from 6 to 40 weeks. Following initial vacuuming to remove surface contamination, individual carpet specimens were removed and placed in a vented, disinfected container at ambient conditions for test durations of 18–36 weeks. They suggested the following assay procedure: Divide the carpet schematically into 100 equal areas. Cut 2.25 cm diameter plugs from eight randomly selected areas. Cut each plug into six pieces using sterile scissors. Macerate two lots of four plugs in 200 ml phosphate buffered water (pH 7.2) using individual Waring blenders and then pool the contents. Prepare ten-fold dilutions and plate onto appropriate media. Assay bacterial counts at Time 0 and at required intervals throughout the test duration.

Agitation Methods

(i) *Agitation procedures for monitoring microbial survival on fabrics*
Inoculate 100 ml portions of tryptic soy broth (Difco) with *Escherichia coli* ATCC 25922 or *Staphylococcus aureus* ATCC 25923 and incubate at 35°C for 18 h. Dilute

with phosphate-buffered saline (0.01 M, pH 7.4) to produce high (10^5 cfu/ml) counts or low (60–600 cfu/ml) counts as appropriate. Cut square swatches of test fabric (approx 1000 cm^2) and autoclave. Seed each swatch with 10 ml of bacterial suspension dispensed dropwise from a sterile pipette. For bacterial recovery at Time 0, place the test swatch in 1 litre of phosphate-buffered saline containing 5% tryptic soy broth in a sterile paint can. Cody *et al.* (1984) proceeded to shake the paint can for 90 s at 580 oscillations per minute using a commercial paint shaker (e.g. Red Devil Model 5100, Red Devil Inc, Union, N.J., U.S.A.). A comparable process may be utilized. Prepare serial dilutions using phosphate-buffered broth. Use membrane filtration or conventional plating techniques to assay bacterial populations. Incubate tryptic soy agar with 5% sheep blood plates at 35°C and count after 24 and 48 h. To assess bacterial survival, repeat the process at suitable time intervals using high and low inoculation levels.

(ii) *Agitation procedures for monitoring microbial survival on carpets*
Expose a test carpet in a heavily trafficked area for a suitable period. Lift the carpet and cut into 15.2 cm squares with a sterile blade. Wipe the backing of the carpet squares with propanol. From each test square, cut twelve 2.5 cm squares and place into sterile bottles containing 25 ml sterile distilled water and a quantity of glass beads. Shake each bottle vigorously by hand for 5 minutes. Prepare serial dilutions and plate out as appropriate (Laskaris 1966). Store the remaining pieces under suitable conditions, and monitor microbial survival at regular time intervals.

(iii) *Agitation of test fabrics in bacterial suspensions*
Prepare cell suspensions of approx 10^4 cells per ml in peptone water (Difco) using 0.095% w/v for *Staphylococcus aureus*, 0.15% w/v for *Escherichia coli* and 0.2% w/v for *Pseudomonas aeruginosa*. Prepare a test system consisting of a polystyrene vial (21.5 mm diameter) containing a 2.5 ml suspension of one of the indicator organisms and a sterile magnetic flea. Mount the test fabric on a glass disc (1.23 cm^2) attached to a glass rod. A rubber sample rod holder is used to position the glass rod above the liquid and to seal the top of the vial. The glass rod bearing the fabric is lowered into the cell suspension at Time 0. The suspension is constantly mixed at 100 rev/min and the temperature maintained at 37°C. Remove 0.1 ml aliquots at regular intervals throughout the experiment, and add to 0.9 ml of buffer or neutralizer (e.g. Tween–Lecithin for fabrics containing chlorhexidine). Plate out 0.2 ml aliquots onto Isosensitest agar (Oxoid) and incubate for 24 h at 37°C. Holland & Davis (1985) have used the method to assess the bactericidal properties of wound dressings using three suggested challenge organisms.

(iv) *Agitation by the use of ultrasonic energy to monitor bacterial survival on carpets*
Inoculate individual sterile bottles containing 150 ml of nutrient broth (Difco) with Bactrol discs (Difco) for *Escherichia coli* ATCC 25922, *Staphylococcus epidermidis*

ATCC 12228 and *Streptococcus faecalis* ATCC 19433 and incubate at 35°C for 48 h on a rotary shaker at 120 rev/min. Prepare three 29.7 ml aliquots of 10% v/v horse serum (Difco) in peptone water in 100 ml autoclavable plastic bottles. Prepare a 1% suspension of the individual 48 h cultures using horse serum/peptone water broth. Cut out thirty 5.0 cm diameter plugs using a sterile carpet cutter at random from a 60 × 30 cm test carpet sample. Stand the plugs (three rows of ten plugs) on aluminium foil on a flat surface, exposed to normal ambient conditions. Inoculate each row of ten plugs with one of the appropriate test organisms using 2 ml of the bacterial suspension. Assay bacterial survival as follows: Dissolve 34 g of potassium dihydrogen phosphate in 500 ml of distilled water and adjust to pH 7.2 with sodium hydroxide (BDH Chemicals) and make up to 1 litre. Prepare a working rinse solution by adding 1.25 ml of the stock phosphate buffer to a 0.02% v/v solution of Tween 80 in distilled water and autoclaving. For Time 0 readings, place one plug for each test organism into separate 250 ml autoclavable plastic bottles containing 30 ml of rinse solution. Place the bottles in an ultrasonic tank (Kerry Ultrasonics Ltd, Herts., U.K.) containing 2 l of freshly made up 0.3% v/v Tween 80 solution. Sonicate the samples at 25 kHz for 5 min. Ensure that the level in the tank is above that in the bottles. Prepare serial dilutions and plate duplicate 0.2 ml aliquots onto MacConkey agar (*E. coli*), Mannitol Salt agar (*S. epidermidis*), and KF Streptococcus agar (*S. faecalis*). Incubate at 35°C for 24 h (MacConkey agar) or 48 h (Mannitol Salt and KF Streptococcus agar). Taking further plugs, repeat the process at suitable time intervals (e.g. 6, 24, 26, 30, 48, 50, 52 and 54 h). Maintain bacterial suspensions in horse serum/peptone water at ambient temperature and plate out appropriate serial dilutions as controls at corresponding time intervals. Express results as colony forming units per carpet plug (or ml for the controls). The method may be used to compare simultaneous bacterial survival rates on different carpet types (see Figs 2a and b).

Agar Overlay Methods

(i) *Agar overlay—method for monitoring survival of bacteria on fabrics*
From a pre-seeded or exposed test substrate, aseptically cut a disc specimen from the bulk test cloth. Irradiate one side of the disc with ultraviolet light for 30 s with the specimen placed 15.2 cm from a mercury vapour lamp or similar source. Place the specimen, irradiated side down, in a sterile petri dish. Overlay with 25 ml nutrient agar (Difco) and incubate for 48 h at 35°C. Count the resultant colonies (Charnley & Eftekar, 1969). Remove and test subsequent specimens as required to plot bacterial survival.

(ii) *Agar overlay—triphenyltetrazolium chloride methods*
Prepare 10^5 cfu/ml suspensions of *Escherichia coli* (ATCC 25922) and *Staphylococcus aureus* (ATCC 25923) in phosphate-buffered saline (0.01 M, pH 7.0). Inoculate the

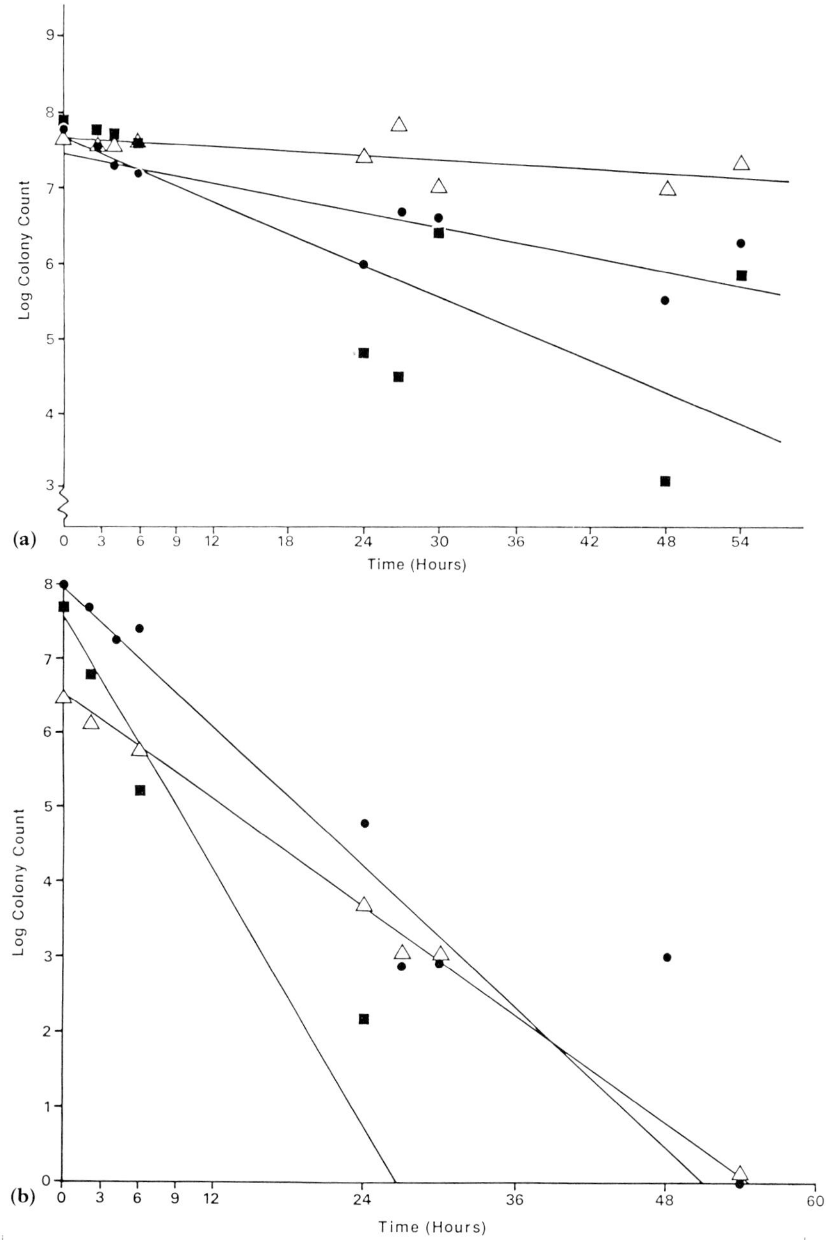

FIG. 2. Survival rates of two test organisms on two separate carpet samples. (**a**) Carpet sample exhibiting little inhibition of bacterial growth. Test organisms grown in nutrient broth, 1% cell suspension prepared in 10% horse serum/peptone water. 2 ml aliquots pipetted onto carpet samples. Survival of test organisms monitored by plate counts. △ *Streptococcus faecalis*, ● *Staphilococcus epidermidis*, ■ *Escherichia coli*. (**b**) Carpet sample exhibiting strong inhibition and rapid die-away of bacterial growth. Protocol and symbols identical.

test specimens. Place the pre-seeded swatches (50 cm^2) in individual petri dishes. Overlay with 20 ml of tryptone glucose extract agar (Difco) containing 15 mg of triphenyltetrazolium chloride (B.D.H.) per litre. Allow to solidify and incubate at 35°C for 24 h. Count the number of red colonies visible on the test swatches (Cody *et al.* 1984). An alternative method involves overlaying the test fabrics with tryptone glucose extract agar only, followed by incubation. Flood the resultant plates with a solution containing 0.1% TTC, 1% glucose and 1% disodium EDTA. Leave for 1 h to allow for diffusion and count the resultant colonies. Note that heavily contaminated fabrics will result in confluent growth.

General Conclusions

A variety of test procedures have been presented for the determination of microbial bioburden on fabrics and carpets. The method selected for individual substrate testing will reflect the nature of the substrate (i.e. carpet or hospital fabric), the degree of information required (i.e. qualitative or quantitative), testing limitations (i.e. destructive or non-destructive) and the purpose of assessment (i.e. direct contamination assay, challenge testing, or monitoring of natural contamination accumulation and/or survival).

References

BLOWERS, R. & WALLACE, K. R. 1955 The sterilization of blankets with cetyl trimethylamine bromide. *Lancet* **i**, 1250–1252.

CHARNLEY, J. & EFTEKAR, N. 1969 Penetration of gown material by organisms from the surgeon's body. *Lancet* **i**, 172–174.

CODY, H. J., SMITH, P. F., BLASER, M. J, LAFORCE, F. M. & WANG, W-L. L. 1984 Comparison of methods of recovery of *Escherichia coli* and *Staphylococcus aureus* from seeded laundry fabrics. *Applied and Environmental Microbiology* **47**, 965–970.

COLBECK, J. C. 1956 Studies in hospital infections. 1. The importance of fomites in the spread of staphylococcal infections with particular reference to mattresses and washing facilities. *Canadian Services Medical Journal* **12**, 563–580.

HOLLAND, K. T. & DAVIS, W. 1985 A note on an in vitro test system to compare the bactericidal properties of wound dressings. *Journal of Applied Bacteriology* **59**, 61–63.

LASKARIS, T. 1966 For what is carpet disinfection needed? *The Modern Hospital* **107**, 132–140.

LEMON, H. M. 1943 A method for collection of bacteria from air and textiles. *Proceedings of the Society for Experimental Biology and Medicine* **54**, 298–301.

LITSKY, B. Y. 1973 Bacteriological techniques for testing carpets. *Health Laboratory Science* **10**, 28–38.

McCARTHY, B. J. 1984 Preservatives for use in the wool textile industry. S.A.B. Demonstration Meeting, October 1984.

NICHOLES, P. S. 1970 Bacteria in laundered fabrics. *American Journal of Public Health* **60**, 2175–2180.

PILSWORTH, R. 1969 Carpet in hospitals. *British Hospital Journal and Social Services Review* 922–926.

PUCK, T. T., ROBERTSON, O. H., WISE, H., LOOSLI, C. G. & LEMON, H. M. 1946 The oil treatment of bedclothes for the control of dust-borne infection. 1. Principles underlying the development and use of a satisfactory oil-in-water emulsion. *American Journal of Hygiene* **43**, 91–104.

RUBBO, S. D. & DIXSON, S. 1960 A contact-plate technique for determining bacterial contamination of fabrics. *Lancet* **ii**, 394–397.

SCHAFFER, J. G. 1966 Microbiology of hospital carpeting. *Health Laboratory Science* **3**, 73–85.

SUZUKI, A., NAMBA, Y., MATSUURA, M. & HORISAWA, A. 1984 Bacterial contamination of floors and other surfaces in operating rooms. *Journal of Hygiene (Cambridge)* **93**, 559–566.

TEN CATE, L. 1965 A note on a simple and rapid method of bacteriological sampling by means of agar sausages. *Journal of Applied Bacteriology* **28**, 221–223.

WALTER, W. G. & STOBER, A. 1967 Quantitative and qualitative microbial studies of hospital carpets. *Journal of Environmental Health* **30**, 293–300.

WALTER, W. G. & STOBER, A. 1968 Probe method for the microbial sampling of hospital carpets. *Health Laboratory Science* **5**, 162–170.

WETZLER, T. F. 1971 Critical analysis of the microflora of towelling. *American Journal of Public Health* **61**, 376–393.

WILKOFF, L. J., WESTBROOK, L. & DIXON, G. J. 1969a Factors affecting the persistence of *Staphylococcus aureus* on fabrics. *Applied Microbiology* **17**, 268–274.

WILKOFF, L. J., WESTBROOK L. & DIXON, G. J. 1969b Persistence of *Salmonella typhimurium* on fabrics. *Applied Microbiology* **18**, 256–261.

Biocide Tests on Algal and Fungal Growths on Solid Substrate

L. H. G. Morton
School of Applied Biology, Lancashire Polytechnic, Preston, Lancs. PR1 2TQ

Algae and fungi are known to colonize the external surfaces of buildings and monuments (Perrichet 1983), giving the surface a dirty, neglected and unsightly appearance, but, in addition, they are considered to be the fore-runners of lichens, mosses and higher plants capable of extensive corrosive activity.

In terrestrial environments epiphytic algal growths will occur on surfaces where conditions of dampness, warmth and light are conducive (Hueck-van der Plas 1968; Whiteley 1973; Richardson 1973; Springle 1975; Morton 1979), and fungi can occur on stonework as independent colonizers or as lichen symbionts (Grant & Bravery 1981a; Perrichet 1983).

A number of methods are available for assessing the efficiency of compounds intended for use as algicides. Some of these have been reviewed (Grant 1982) and evaluated (Grant & Bravery 1981a). Interactions which may occur between organism/substrate/biocide are difficult to assess other than by field trials. However, Grant & Bravery (1981a, b) describe a vermiculite bed technique for the evaluation of algicides for use with constructional materials, including mortar and stone blocks. By using an additional substrate, this test can be modified to be suitable for the testing of fungicides. This substrate is a calcium silicate building board (interior grade) reinforced with a small quantity of vermiculite and containing a cellulose wood fibre matrix (trade name 'Masterboard'). It contains no asbestos or man-made fibre. Because of the cellulose content of the board it is a suitable substrate for growing cellulolytic test fungi. Morton (1979) describes a technique for isolating and culturing algae from stonework. Cultures obtained in this manner, often mixed cultures of more than one algal species, can be used in the tests described below, although selected monocultures can also be used. Cellulolytic dematiaceous hyphomycetes are suitable fungal test organisms.

Test Procedures

Materials required for tests

(1) Transparent perspex boxes measuring 205 × 96 × 75 mm with lids.

Industrial
Microbiological Testing

0-632-01793-7

(2) Vermiculite (Dupré vermiculite grade DSF). (3) Deionized water. (4) 95% ethyl alcohol. (5) Board test pieces measuring approximately $90 \times 30 \times 5$ mm (trade name 'Masterboard'). (6) Sterilizable plastic bags measuring 40×60 cm (trade name 'Sterilin'). (7) Supply of CO_2. (8) Sterile nutrient salts solution: Medium C (Kratz & Myers 1955). This medium has the following composition: $MgSO_4.7H_2O$, 0.25; K_2HOP_4, 1.00; $Ca(NO_3).4H_2O$, 0.25; KNO_3, 1.00; Na_3 citrate. $2H_2O$, 0.165; $NaHCO_3$, 1.00; $Fe_2(SO_4)_3.6H_2O$, 0.004; microelements solution, 1 cm^3. The microelements solution has the following composition (g/l): H_3BO_3, 2.86; $MnCl_2.4H_2O$, 1.81; $ZnSO_4.7H_2O$, 0.222; MoO_3, 0.01777; $CuSO_4.5H_2O$, 0.079. (9) 'Mixed' algal suspension (Morton 1979) and/or selected algal monocultures (Morton *et al.* 1984). (10) Spore suspension of fungal test organisms. (11) Glass spreader. (12) Small soft haired paint brush. (13) Biocide as aqueous solution. (14) Pasteur pipettes. (15) Malt agar (Oxoid Ltd). (16) Illuminated incubator (e.g. a Gallenkamp orbital illuminated incubator is preferable but not essential.)

To obtain algal growth on 'Masterboard' test pieces

In the vermiculite test (Grant & Bravery 1981a) each box is charged with 100 g vermiculite. In this modification of the test the vermiculite is autoclaved in 100 g aliquots in separate 'Sterilin' bags at 121°C for 20 min and the interior of the perspex boxes sterilized by alcohol-wiping prior to use. Sterile deionized water is then added to each box, sufficient to thoroughly dampen the vermiculite without it becoming mobile (350–400 ml water is sufficient). The 'Masterboard' test pieces are also autoclaved in plastic bags into which, when cool, CO_2 gas is introduced. The bags are sealed using an elastic band and the test pieces are kept in this atmosphere for 3 weeks. The test pieces are then removed and placed onto the moist vermiculite in the perspex boxes. No more than four test pieces per box are recommended and they should be allowed to equilibrate for 48 h with the lids of the boxes firmly in place. The test pieces are then surface spread with 0.2 ml of algal suspension and the boxes are placed in an illuminated incubator set at 25°C. Alternatively the boxes can be placed in direct sunlight on a window ledge. The lids of the boxes should be firmly in place.

Sparse but uniform growth should occur on the test pieces in 7 to 10 days. This growth can be enhanced by the application of 0.2–0.4 cm^3 of the sterile nutrient salts solution onto the surface of the test pieces at twice-weekly intervals for a further 2–3 weeks.

To obtain fungal growth on 'Masterboard' test pieces

Test pieces are autoclaved in a 0.5% peptone solution (Oxoid Ltd) for 20 min at 121°C. The test pieces are allowed to cool in the solution for 12 h, after which they are placed in the perspex boxes and inoculated by surface spreading with a spore suspension of the test fungus. The spore suspension is prepared by adding 5 cm^3 of sterile distilled water to a sporing culture of a suitable fungus, e.g. *Ulocladium atrum* or *Stachybotrys chartarum*, and pipetting the suspension into a universal bottle. An inoculum of 0.2 ml is used. A fungal lawn will develop on the test pieces after 5 days incubation at 25°C.

Application of biocide solution to test pieces

Biocide solutions of known concentration may be applied to one half of the microbial growth on the test pieces using a small soft paint brush (Morton & Coupe 1985).

Interpretation of results

The effect of a biocide upon algal growth can be seen within a few days (sometimes hours) of application, in that the green 'lawn' disappears or becomes discoloured where the biocide has been in contact with the cells (Fig. 1). Since dematiaceous fungi produce dark or black spores (Fig. 2), viability of treated fungal growth can be assessed by pressing the board surface onto the surface of malt agar plates and examining the plates for growth after suitable incubation.

Acknowledgments

I am grateful to Mr C. Grant and Dr A. F. Bravery of the Building Research Establishment and to Mr R. Springle of the Paint Research Institute for their interest in this work.

FIG. 1. From left to right: 'Masterboard' uninoculated; 'Masterboard' with algal lawn; 'Masterboard' half treated with a biocide.

FIG. 2. Stages in the colonization of 'Masterboard' by a fungus producing black spores.

References

GRANT, C. & BRAVERY, A. F. 1981a Laboratory evaluation of algicidal biocides for construction materials. 1. An assessment of some current test methods. *International Biodeterioration Bulletin* **17**, 113–123.

GRANT, C. & BRAVERY, A. F. 1981b Laboratory evaluation of algicidal biocides for constructional materials. 2. Use of vermiculite bed technique to evaluate a quaternary ammonium biocide. *International Biodeterioration Bulletin* **17**, 125–131.

GRANT, C. 1982 Fouling of terrestrial substrates of algae and implications for control. *International Biodeterioration Bulletin* **18**, 57–65.

HUEK VAN DER PLAS, E. H. 1968 The microbiological deterioration of porous building materials. *International Biodeterioration Bulletin* **4**, 11–28.

KRATZ, W. A. & MYERS, J. 1955 Nutrition and growth of several blue green algae. *American Journal of Botany* **42**, 282–287.

MORTON, L. H. G. 1979 Laboratory test methods for assessing the algicidal properties of biocidal wash preparations. *International Biodeterioration Bulletin* **15**, 84–86.

MORTON, L. H. G. & COUPE, T. S. 1985 Report on vermiculite test bed system using algal cultures. *International Biodeterioration Research Group Document* No. C85/01.

MORTON, L. H. G., LOVELL, M. L. & MITCHELL, A. F. 1984 The effect of maceration on filamentous algae used for the testing of algicidal compounds. *International Biodeterioration* **20**, 33–36.

PERRICHET, A. 1983 The growth of microorganisms on concretes and renderings. RILEM *Technical Committee* **62** 'Soiling and cleaning of facades', 18pp.

RICHARDSON, B. A. 1973 Control of biological growths. *Stone Industries* **8**, 2–6.

SPRINGLE, R. 1975 Testing biocidal paints. *Some Methods for Microbiological Assay. Society for Applied Bacteriology Technical Series* No. 8, pp. 191–202.

WHITELEY, P. 1973 The occurrence and control of moulds and algae on paints. *Journal of the Oil Colorists Chemical Association* **56**, 382–387.

Biocide Tests on Algal Suspensions Isolated from Stonework

L. H. G. Morton
School of Applied Biology, Lancashire Polytechnic, Preston, Lancs. PR1 2TQ

Studies of epiphytic algal growth on stonework and other exterior surfaces by European researchers (Dukes 1972; Grant 1982; Perrichet 1983; Ryan 1983) and Asian researchers (Young *et al.* 1972; Wee & Lee 1980) have shown that smooth surfaces are less affected than rough cement based surfaces with high porosity.

Algae are considered to be the fore-runners of lichens and mosses capable of more extensive damage, and hence a wide range of biocidal treatments for fouled surfaces has been proposed (Bravery 1977, 1981; Springle 1979).

The development of laboratory methods for screening compounds for use as algicides has arisen because of the time scale and practical implications associated with field trials. Many of these tests have been reviewed (Grant 1982; Morton 1986) and evaluated (Grant & Bravery 1981a, b). These techniques often use pure cultures of unicellular algae. It has been suggested (Morton 1979; Morton *et al.* 1984) that mixed cultures of algae isolated from fouled stonework, representative of the flora of that environment, should be used in tests which assess the effectiveness of compounds intended for use as algicides. Algal growth from stonework can be obtained using the technique described below. Some of the isolates from stonework may be filamentous and should only be used as test organisms if uniform suspensions can be obtained which are viable. The effect of maceration upon the viability of filamentous algae has been studied (Morton *et al.* 1984).

Experimental Procedures

Materials required

(1) Alcohol (for flame-sterilizing). (2) Scalpel or small scraper. (3) Sterile distilled water (in universal bottles). (4) 250 ml conical flasks: plugged and sterile. (5) Sterile plugged test tubes. (6) Sterile 1 ml pipettes. (7) 100 ml measuring cylinder. (8) Petri dishes (plastic). (9) Agar (Oxoid Ltd). (10) Medium C (Kratz & Myers 1955). (11) Algal broth (Difco Ltd). (12) Solution of benzyl penicillin G (Benzylpenicillin SIGMA Stock No. PEN-NA). (13) Biocide solution(s) under test. Filter paper discs

Industrial
Microbiological Testing
0–632–01793–7

0.5 cm diameter. (14) Mixed algal suspension (Morton 1979). (15) Cultures of filamentous algae, e.g. *Oscillatoria tenuis* and *Trentepohlia odorata*. (16) Illuminated incubator. (17) Illuminated orbital incubator. (18) Device for macerating/breaking up algal filaments, e.g. FISONS W.M./250/F WHIRLIMIXER or POTTER ELVEJEM homogenizer or Ultrasonic bath (L & R Manufacturing Company, Wealdstone, Harrow).

Preparation of bacteria-free algal suspension from fouled stonework (Morton 1979)

Culture medium. The medium is Medium C of Kratz & Myers (1955) and has the following composition (g/l): $MgSO_4.7H_2O$, 0.25; K_2HPO_4, 1.00; $Ca(NO_3).4H_2O$, 0.025; KNO_3, 1.00; Na_3 citrate. $2H_2O$, 0.165; $NaHCO_3$, 1.00; $Fe_2(SO_4)_3.6H_2O$, 0.004; Microelements solution, 1 cm^3.

The microelements solution has the following composition (g/l): H_3BO_3, 2.86; $MnCl_2.4H_2O$, 1.81; $ZnSO_4.7H_2O$, 0.222; MoO_3, 0.01777; $CuSO_4.5H_2O$, 0.079.

The medium can conveniently be made up by using the following concentrated stock solution to make up 1 litre: $MgSO_4.7H_2O$, 50 g in 200 ml (add 1 ml); KH_2PO_4, 125 g in 500 ml (add 4 ml); $CaNO_3.4H_2O$, 5 g in 200 ml (add 1 ml); Na_3 citrate. $2H_2O$, 33 g, and $Fe_2(S)_4.6H_2O$, 0.08 g, together in 200 ml (add 1 ml).

In order to suppress non-algal growths in this medium, penicillin G is added at the rate of 12 units/ml. The medium is first sterilized by autoclaving at 121°C for 20 min. When cool the penicillin is added. The penicillin solution may be sterilized by filtration before it is added to the medium but this has not been found to be essential.

Mixed algal culture. Epiphytic algal growth on stonework is removed by scraping the surface with a flamed scalpel. The algal 'dust' obtained in this manner is suspended in 10 ml of sterile distilled water contained in a universal bottle. This crude algal suspension is inoculated directly into 150 ml culture medium containing 12 units/ml of penicillin and held in a conical flask. The flask is placed in an illuminated orbital shaker and incubated at 25°C for 7–10 days by which time a mixed culture of algae should be available for assay purposes.

A shaker which has been found suitable for this test is made by Gallenkamp; it has seven fluorescent tubes, of 30 W each, mounted 350 mm above the flasks. Alternatively, the test will work quite well if flasks are placed in direct sunlight on a window ledge. However, this is only successful in summer. The flasks must be shaken daily.

Determination of toxic level (lethal dose) of biocide

By the addition of increasing quantities of biocide to fixed volumes of algal suspension. In this method quantities of the biocidal preparation ranging from 0.05 to 5.0 cm^3 are introduced into sterile tubes containing 5.0 cm^3 of algal suspension. The volume of biocide to be added will depend on its concentration and effectiveness, but two suitable working ranges of biocide volume have been found to be:

Range 1 0.05 0.1 0.15 0.2 0.25 0.3 0.35 0.4 0.45 0.5.
Range 2 1.0 1.5 2.0 2.5 3.0 3.5 4.0 4.5 5.0.

The volume in each tube is made up to 10 cm^3 with sterile distilled water where necessary so that each viability assessment can be made in the same volume of liquid. The tubes are illuminated and held at room temperature for 10 days. The criterion for assessing the viability of the algae is the persistence or otherwise of green pigmentation.

By the introduction of a standard volume of algal suspension into solutions containing varying amounts of growth medium and biocide solution. In this method 10 ml aliquots of algal suspension are introduced into flasks each containing a standard volume of solution made up of varying amounts of growth medium and biocide solution. The flasks are incubated in the orbital shaker with illumination for 1 week at 25°C. Viability of the algal test organism is assessed using the pigmentation criterion. A suitable initial working range of dilutions is given in Table 1.

Confirmation of algal death—the criterion of reoccurrence of pigmentation
Algal death can be confirmed by collecting algal debris resulting from the tests. The debris should be washed using decanting and centrifuging techniques and then introduced into flasks containing sterile 100 ml volumes of culture medium. The flasks should be incubated at 25°C in the illuminated orbital shaker, or at room temperature in the light, for 2 weeks. Death of the algae is confirmed when there is no re-occurrence of pigmentation.

Determination of the effect of biocide upon macerated filamentous algae (Morton *et al.* 1984)

Assessment of viability of macerated material. Liquid shake cultures of the filamentous test algae are grown in algal broth. Stock cultures are maintained on this same medium containing agar (20 g/l). Shake-culturing is carried out in 250 ml flasks containing 100 ml of algal broth. Aliquots of algal suspension should be

TABLE 1. *Suitable range of biocide concentration for determination of biocide toxicity*

Flask	Volume of culture medium (ml)	Volume of biocide (ml)	Volume of algal suspension (ml)
A	90	10	10
B	80	20	10
C	60	40	10
D	40	60	10
E	20	80	10
F	100	0	10

macerated until filaments are no longer visible to the naked eye, i.e. a uniform green suspension is obtained. 0.2 ml of suspension is spread onto the surface of algal broth agar plates. The plates, when dry, should be inverted and incubated with illumination for 7–14 days. This procedure should be repeated at daily intervals in order to establish the effect of maceration upon the viability of test organisms.

Assessment of algicidal activity. This can be undertaken in one or other of three ways:

(i) Using a modification of the technique of Wright (1975), 0.1 ml of macerated algal material is spread onto the surface of algal broth agar and allowed to dry. Biocide solutions of known concentration are applied to filter paper discs measuring 0.5 cm in diameter. The discs, after blotting to remove surplus biocide, are placed onto the seeded plates. The plates are incubated with illumination at 25°C for up to 14 days. Zones of inhibition form around the discs containing biocide (Fig. 1).

(ii) Six separate drops of a heavy algal suspension are inoculated onto the surface of a plate containing algal growth agar and are allowed to dry. Filter paper discs containing biocide solutions of known concentration, together with controls containing no biocide, are placed 2 mm from the margins of the drop-inocula. The plates are illuminated at 25°C with illumination for up to 14 days. Distinct areas of inhibition become visible as incubation proceeds (Fig. 2).

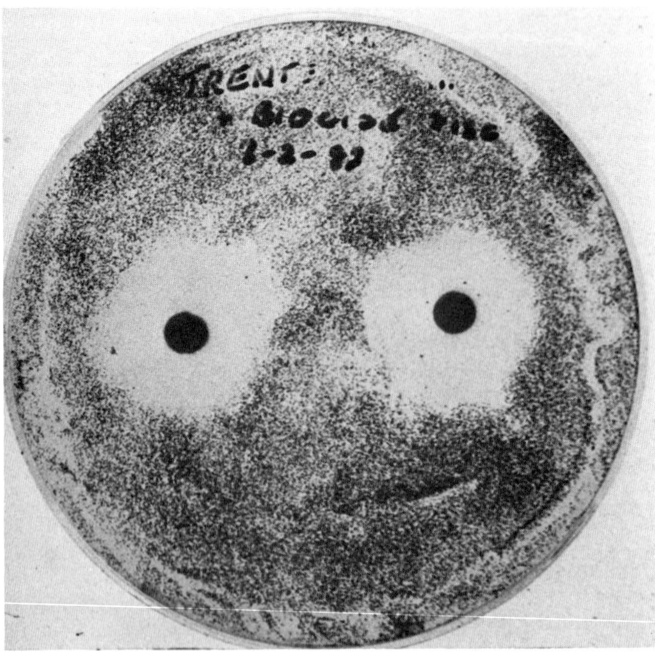

FIG. 1. Algal lawn of *Trentepohlia odorata* with zones of inhibition around filter paper discs impregnated with a biocide.

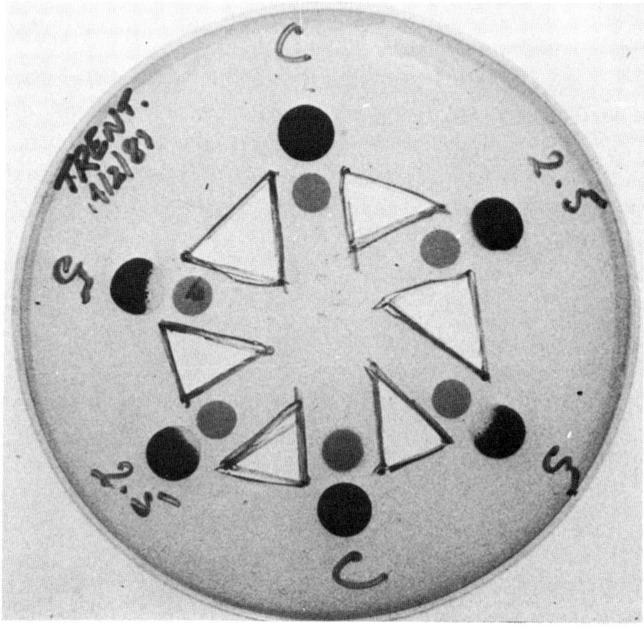

FIG. 2. Inhibition areas in drop-inocula of algal suspension.

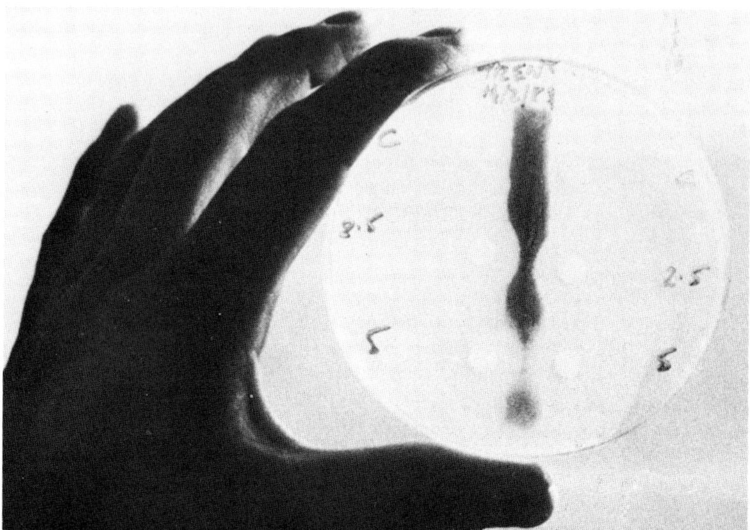

FIG. 3. Inhibition of a 'strip-inoculum' of algal suspension.

(iii) A band or strip of algal suspension measuring approximately 1.0×7.0 cm is carefully inoculated onto an agar plate of algal growth medium. This can be achieved by placing a piece of paper with the area to be inoculated marked upon it under the agar plate and spreading the algal suspension carefully over the agar using a Pasteur pipette. Impregnated discs and control discs are placed on either side of the band 2 mm from its margin. The plates are incubated with illumination as before. Distinct areas of inhibition of algal growth become visible as incubation proceeds (Fig. 3).

Acknowledgment

I am grateful to Durham Chemicals Ltd, who provided the biocidal compounds used in the development of these tests.

References

BRAVERY, A. F. 1977 Compounds associated with the eradication and prevention of biological growths on constructional materials. *International Biodeterioration Research Group* Document No. C/77/03.

BRAVERY, A. F. 1981 Preservation in the construction industry. In: *Principles and Practice of Disinfection Preservation and Sterilization*, eds Russell, A. D., Hugo, W. B. & Ayliffe, G. A. J. Oxford: Blackwell Scientific Publications.

DUKES, W. H. 1972 Conservation of stone: Causes of decay. *Architects Journal* **156**, 422–429.

GRANT, C. 1982 Fouling of terrestrial substrates by algae and implications for control. *International Biodeterioration Bulletin* **18**, 57–65.

GRANT, C. & BRAVERY, A. F. 1981a Laboratory evaluation of algicidal biocides for use on constructional materials. 1. An assessment of some current test methods. *International Biodeterioration Bulletin* **17**, 113–124.

GRANT, C. & BRAVERY, A. F. 1981b Laboratory evaluations of algicidal biocides for constructional materials. 2. Use of vermiculite bed technique to evaluate a quaternary ammonium biocide. *International Biodeterioration Bulletin* **17**, 125–131.

KRATZ, W. A. & MYERS, J. 1955 Nutrition and growth of several blue green algae. *American Journal of Botany* **42**, 282–287.

MORTON, L. H. G. 1979 Laboratory test methods for assessing the algicidal properties of biocidal wash preparations. *International Biodeterioration Bulletin* **15**, 84–86.

MORTON, L. H. G. 1986 A review of techniques available for testing algicides for use in terrestrial and fresh water environments. *International Biodeterioration* **22**, 5–9.

MORTON, L. H. G., LOVELL, M. L. & MITCHELL, A. F. 1984 The effect of maceration on filamentous algae used for the testing of algicidal compounds. *International Biodeterioration* **20**, 33–36.

PERRICHET, A. 1983 The growth of microorganisms on concretes and renderings. RILEM *Technical Committee* 62 'Soiling and cleaning of facades'. 18pp.

RYAN, N. M. 1983 Algal growth on cementitious surfaces. Pilot study, National Institute for Physical Planning and Construction Research, Dublin.

SPRINGLE, R. 1979 Testing biocidal paints. *Some Methods for Microbiological Assay. The Society for Applied Bacteriology Technical Series No. 8*, pp. 191–202.

Wee, Y. C. & Lee, K. B. 1980 Proliferation of algae on surfaces of buildings in Singapore. *International Biodeterioration Bulletin* **16**, 113–117.

Wright, S. J. L. 1975 Use of micro-algae for the assay of biocides. In: *Some Methods for Microbiological Assay*, eds Board, R. G. & Lovelock, D. W. London: Academic Press.

Yong, F. N., Yeow, C. T., Chua, N. H. & Wong, H. A. 1972 Method for screening and evaluating algicidal and algistatic surface coatings. *Journal of the Singapore Institute of Architects* **53**, 13–19.

Bacterial Barrier Properties of Textiles

D. S. Wales, P. F. Hamlyn and B. F. Sagar
Shirley Institute, Didsbury, Manchester M20 8RX

It has been estimated that, in the U.S.A., some 1.5 million of the 20 million patients admitted to hospital annually for surgical procedures fall victim to post-operative infection (Seaman 1980). Such infection may result from a variety of causes, such as faulty aseptic techniques or high microbial load, and in some cases will be unavoidable. One important source of post-operative infection is the passage of bacteria from the skin of the surgeon through the various materials used as gowns and drapes, to the patient.

There are two main ways of preventing, or at least reducing, the passage of bacteria through a fabric. The first approach is to design a material that, due to its construction or inherent properties, will restrict the number of bacteria able to pass through it. This objective is achieved in some materials by incorporating a microporous plastic film between two layers of fabric (Laufman *et al.* 1980), in which the pores are too small to allow the passage of bacteria or liquid water, but are still large enough to allow the fabric to be breathable, i.e. to allow the passage of water vapour. Materials of this design are usually total barriers. However, the passage of bacteria through other types of material, for example nonwoven materials, is often reduced when compared to passage through woven cotton. The finish applied to such fabrics may have an important effect on their bacterial properties. For example, a water-repellent finish will limit the penetration of the material by aqueous fluids, thereby reducing the passage of any contaminating bacteria. Indeed, Olderman (1984) suggests the measurements of liquid repellency may be used to assess surgical barrier materials.

A second method of reducing the passage of bacteria through a fabric, perhaps more relevant to operating drapes than gowns, is to treat the fabric with an anti-bacterial finish. However, the drawback of this method is that the bacteria must remain in contact with the fabric long enough for the biocide to be effective.

Methods of Assessment

Bacterial barrier measurement

Several methods have been suggested for the determination of the bacterial barrier

Industrial
Microbiological Testing
0–632–01793–7

properties of fabrics. There has been some discussion as to whether aseptic barrier performance can be assessed using a test method that does not utilize a bacterial challenge (Olderman 1984). However, since it is possible to obtain materials that will allow the passage of liquid, but not the passage of bacteria (e.g. millipore filters), and even though such materials are not as yet used in barrier fabrics, it would seem necessary to have a method that includes bacteria in the challenge. Another consideration is that purchasers may be sceptical of claims that barrier fabrics possess adequate bacterial barrier properties when such claims are based on tests which do not include the use of bacteria.

A number of methods to determine the bacterial barrier properties of materials have been published and one of the better known is that of Laufman *et al.* (1975). The material under test is supported by metal posts in the form of a hammock, and a bacterial suspension, followed by a weight, is placed on the hammock. The number of bacteria passing through the fabric is then assessed by touching a contact agar plate firmly on the underside of the hammock after 5, 15 and 30 min. Fabrics are rated by determination of the time taken for a positive strike-through of bacteria to be detected. A number of other tests, all using the sensor plate technique, have been reviewed by Marass (1983).

One disadvantage of this type of test method is that the assessment only allows a broadly-based ranking of the effectiveness of fabrics as bacterial barriers. For example, many of the assessment schemes have only four categories: 0 = no breakthrough; 1 = one colony, <25% of contact plate contaminated; 2 = 25–50% of contact plate contaminated; 3 = >50% of contact plate contaminated. Also the force exerted by the operator in touching contact plates to the underside of fabrics, and the localized stress on the fabrics caused by the weights, are sources of variation in such tests. The following test method was developed with the objective of reducing experimental variation, and of achieving more accurate ranking of materials with respect to their bacterial barrier properties.

Shirley Institute aqueous bacterial barrier method

The testing chamber consists of two 250 ml glass vessels clamped together, with a 12.5 cm diameter disc of fabric separating the two halves. Each glass vessel has a glass tube (0.9 cm i.d.) protruding vertically from one side and contains a magnetic stirring bead (Fig. 1). The edge of each disc is sealed with 'Seam Stuff' seam sealer (W. L. Gore, Kirkton Campus, Livingston, Scotland) to prevent wicking during testing, and the disc is sealed into the testing chamber using silicone grease to prevent leakage. After assembly, a rubber bung is placed in the glass tube on the test side of the unit, and the unit is autoclaved at 121°C for 15 min. Sterile saline [0.9% w/v NaCl] is added to both sides of the unit, until the levels reach just below the base of the glass tubes. The rubber bung is replaced in the glass tube on the test side of the unit, and an appropriate amount of a 100-fold dilution of a 16 h culture of *Staphylococcus aureus* (grown in tryptone soya broth, Oxoid Ltd) to give a final

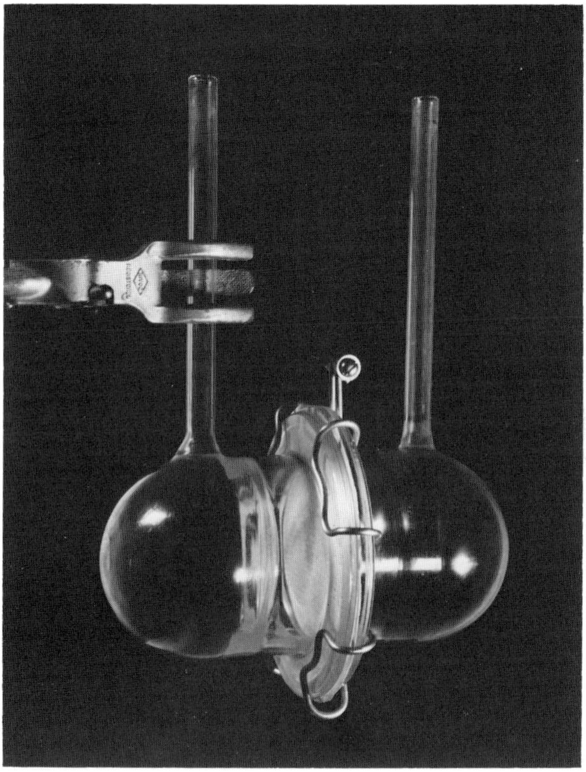

FIG. 1. The Shirley Institute testing chamber.

bacterial concentration of approximately 10^4 cfu ml^{-1} is then added to the challenge side of the unit. The challenge culture is mixed by actuating the magnetic stirring bead for a few revolutions and more saline added until the level is 10 cm above the base of the glass tube. A 0.1 ml sample of the challenge culture is taken, and a plate count undertaken using saline as dilution fluid and tryptone soya agar plates (Oxoid Ltd). The rubber bung is then removed from the test side of the unit, and after 10 min the test saline is mixed by means of the magnetic stirring bead and a sample taken. The number of bacteria that have penetrated the fabric is then determined by the plate count technique.

The percentage transference of bacteria can then be determined.

Some sample results for this test are presented in Table 1. The cotton fabric was a green open weave material traditionally used in operating theatre gowns. The percentage transfer figure of 1.6 for that material is typical of fabrics that offer little resistance to bacterial transmission. The three nonwoven fabrics are better barriers to bacteria than the cotton, nonwoven fabric 3 having the lowest percentage transfer.

TABLE 1. *Bacterial barrier properties of fabrics as assessed by the Shirley Institute method*

Fabric	Bacterial count (cfu ml^{-1}) Challenge	Bacterial count (cfu ml^{-1}) Test	Percentage transfer
Woven cotton	7.4×10^4	1.2×10^3	1.62
Nonwoven 1	3.2×10^4	1.3×10^2	0.41
Nonwoven 2	4.3×10^4	1.2×10^2	0.28
Nonwoven 3	4.1×10^4	7.0×10^1	0.17
Laminate	2.8×10^4	0	0

No bacteria could be detected on the test side of the challenge apparatus when the laminate was used; this material contains a sheet of microporous plastic film sandwiched between two layers of nonwoven material, and is seen to be totally effective as a bacterial barrier.

Although laminates, such as the one tested, have ideal bacterial properties, such materials, despite being 'breathable', are often uncomfortable to wear for long periods of time, the wearer becoming very hot and sticky. Also the materials are more expensive to produce than conventional woven and nonwoven fabrics, and cost becomes an important consideration.

These results demonstrate that this test method enables fabrics to be easily ranked in order of their effectiveness as bacterial barriers. The method is less prone to experimental variation as the pressure forcing the bacteria through the fabric is applied by an easily reproducible 10 cm hydrostatic head, and the results are determined quantitatively by a serial dilution and plate count technique and not by the use of contact plates. However, there will be some experimental variation associated with the serial dilution and agar plate count method of assessment.

Method for fabrics treated with an antibacterial finish

Operating drapes, especially those in areas adjacent to the operating site, may be treated with an antibacterial finish, as an extra precaution against post-operative infection. However, the aqueous bacterial barrier test described above may not truly assess the efficacy of the finish, because, in the apparatus, the test bacteria may pass through and away from the drape too quickly for the biocide in the drape to act. Such a test would not simulate the situation of a drape next to a patient's skin, where bacteria would be confined within the area of the drape. The following test has been designed to assess the efficacy of antibacterial finishes on fabrics.

Circular swatches (approximately 50 mm in diameter) are cut from the fabric. The number of swatches used in each test container is that required to absorb 1 ml of inoculum, leaving no free liquid in the container. The swatches are sterilized, either by autoclaving, or by exposure to ethylene oxide vapour. A control fabric without the antibacterial agent is also required.

Swatches of test and control fabrics are placed in sterile 8 oz screw-topped wide-mouthed glass bottles. The swatches are inoculated with 1 ml of an appropriate dilution of a 24 h bacterial culture (suitable media and culture conditions are described in Anon. 1981) to give $10\text{–}20 \times 10^4$ cfu ml^{-1} as determined by the subsequent plate count. Immediately after inoculation, 100 ml of a suitable neutralizer solution (e.g. Letheen broth, Difco Laboratories) are added to half the control and test fabric swatches, and the jars shaken vigorously for 1 min. Serial dilutions of this neutralizer solution after shaking are then plated out on suitable nutrient agar (Anon. 1981) so that the cfu recoverable after zero contact time can be found. The remaining jars are incubated at 37°C for 18–24 h, and the cfu recoverable after incubation determined as described above.

The cfu recovered from the control fabric after incubation should be higher than the cfu recovered from the control fabric at zero contact time for the results of the test to be valid. Provided this condition is established, the percentage reduction in cfu of the test fabric is given by

$$[1 - 2A/(B + C)] \times 100$$

where A = the cfu recovered from the test fabric after incubation, B = the cfu recovered from the test fabric at zero contact time, and C = the cfu recovered from the control fabric at zero contact time.

If there is an increase in the cfu recovered from the test specimen after incubation, compared with the cfu recovered from the zero contact time specimens, the antibacterial agent is ineffective. No significant difference between these values may indicate that the agent is acting bacteriostatically. A significant reduction in cfu indicates bactericidal activity.

References

ANON. 1981 Antibacterial finishes on fabrics, evaluation of AATCC Test Method 100. *AATCC Technical Manual* **57**, 304–306.

LAUFMAN, H., EUDY, W. W., VANDERNOOT, A. M., LIU, D. & HARRIS, C. A. 1975 Strike-through of moist contamination by woven and nonwoven surgical materials. *Annals of Surgery* **181**, 857–862.

LAUFMAN, H., MONTEFUSCO, C., SIEGAL, J. D. & EDBERG, S. C. 1980 Scanning electron microscopy of moist bacterial strike-through of surgical materials. *Surgical, Gynaecology and Obstetrics* **150**, 165–170.

MARASS, E. B. 1983 Guideline for selection and processing of aseptic barrier material. Association for the Advancement of Medical Instrumentation USA 10/82 revision.

OLDERMAN, G. M. 1984 Liquid repellency of surgical barrier materials. *Journal of Industrial Fabrics* **3**, 30–41.

SEAMAN, R. E. 1980 Properties and performance of operating room gowns. *Nonwovens Industry* **11**, (4), 6–12.

A Method for the Determination of Enzymatic Activity of Microbial Films on Surfaces

R. A. Collett and J. W. Hopton*
Worcester College of Higher Education, Worcester WR2 6AJ, and
**Department of Microbiology, The University, P.O. Box 363, Birmingham B15 2TT*

A substantial proportion of microbial inhabitants of aquatic environments can be attached to surfaces (Geesey *et al.* 1978). These surface films of organisms are active in the overall metabolism of such environments and can exert specific effects, such as fouling or corrosion, on their substrata (Marshall 1976).

The measurement of the metabolic activities of surface films presents problems because low levels of biomass are usually involved and these produce but small changes in reactant or product concentration in feasible assay times. The use of isotopically labelled substrates permits sensitive assessment of general assimilatory activity (Bott & Brock 1970), but where measurement of specific enzymatic dissimilatory activity is required, labelled substrates are of less utility. For such determinations, nitrophenyl and umbelliferyl linked substrates are of value. They allow a variety of dissimilatory systems, such as glycosidases, esterases and phosphatases, to be assayed. The reaction products, nitrophenols or umbelliferones, can be determined, spectrophotometrically or fluorimetrically respectively, with considerable sensitivity. The latter procedure is generally the more sensitive.

The method described here provides a system in which known areas of substrate can be placed in the environment and recovered for convenient assay of the metabolic activities of adherent microbial films.

Equipment and Experimental Procedure

The method is illustrated here using 4-nitrophenyl phosphate (4NPP) and 4-methylumbelliferyl phosphate (4MUP) for the determination of phosphatase activity of microbial films.

Rack for Immersion of Substratum Strips

The design of the rack is shown in Fig. 1A. Rigid clear PVC sheet is generally quite

Industrial
Microbiological Testing
0–632–01793–7

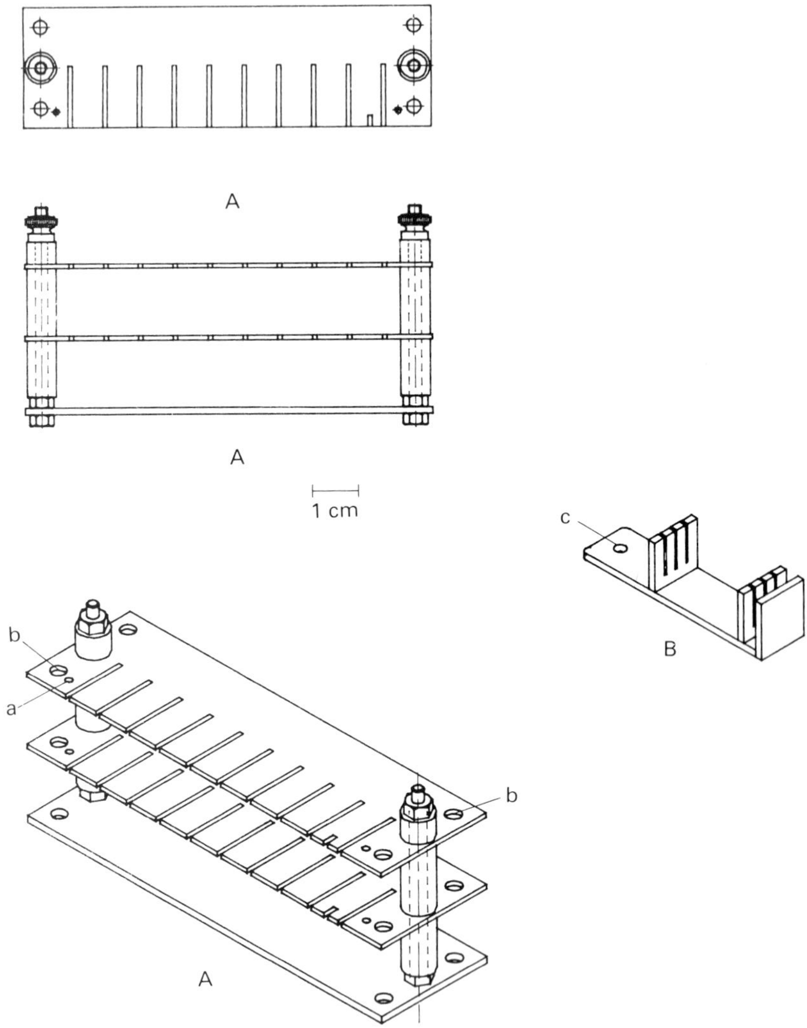

FIG. 1. (A) Rack to accommodate substratum strips for immersion in water. (B) Carriage to accommodate colonized strips for assessment of enzymatic activity. Bar = 1 cm.

adequate for the rectangular plates, but other easily worked rigid plastics have also been used and stainless steel is advantageous if there is a requirement for sterilization. The pillars are cut from threaded stainless steel, brass or plastic rod; nuts of appropriate material are used to secure the plates to the pillars. Separators cut from plastic tubing keep the plates apart. Substratum strips (glass coverslips 9×35 mm Nos. 1 and $1\frac{1}{2}$, Chance Propper Ltd, have been used extensively) are held in the slits in the plates and are kept in place with thin wire (stainless steel, nichrome or plastic covered telephone cable core) anchored through the holes (a) in the plates. The strips become colonized on both sides.

The racks are installed vertically or horizontally in the environment. They are retained in position on plastic base plates equipped with threaded rod pillars which are passed through the holes (b) in the rectangular plates of the carriage. The plastic base plates can be secured with suitable weights or pegs.

The rack and its load of substratum strips are transported back to the laboratory immersed in their environmental water contained in closed plastic boxes. Plastic base plates, with pillars having the same function as those used to retain the carriages in the environment, and of a size sufficient to fit in the base of the box, are used to secure the carriages in transit.

The Assay Procedure

Treatment of colonized substratum strips

The strips are removed from the rack, rinsed in appropriate fluid (the environmental water, sterilized if necessary, is often the most suitable) and then drained by touching the edge of the strip on absorbent paper. The strips are placed in plastic carriages (Fig. 1B) which are designed to carry one, two or three strips, the number used depending on the activity of the film. Glass carriages of similar design and dimensions but constructed from thin glass rod have also been employed but their fragility can be a drawback. A length of suitably inert pliable wire is attached to the carriage through the hole (c) to permit subsequent easy removal or replacement of the carriage from the reaction mixture. It is obviously inadvisable to allow the films to dry out at this stage.

Spectrophotometric determination

The method of spectrophotometric determination employs 4NPP as substrate and is based on the long established method of Bessey *et al.* (1946). This substrate is hydrolysed by phosphatases to yield 4-nitrophenol (4NP).

The mixture consists of 4NPP (3 mmol/l) and appropriate buffer (50 mmol/l). The buffers employed are as follows:

Phosphatases	*Assay pH*	*Buffer system*
Acid	4.0, 5.0	Sodium acetate-acetic acid
Neutral	6.0, 7.0	N-tris (hydroxymethyl)-methyl-2-amino-ethane sulphonic acid (TES)-HCl
Alkaline	8.5	tris (hydroxymethyl)-aminomethane (TRIS)-HCl

In one investigation a mixed buffer containing 25 mmol/l of each component (TTA buffer) has been employed.

Buffer–substrate fluid, sufficient to cover the strips in the rack, is placed in a suitable tube, which may be the spectrophotometer tube itself. For the 'Spectronic' spectrophotometer, 1 × 15 cm tubes are employed and contain 9.0 ml reaction mixture. The tube and contents are equilibrated briefly in a water bath at the assay temperature with the loaded carriage hanging inside the tube and held above the liquid surface by looping the wire over the tube rim. To start the assay the wire is released to allow the carriage to slide gently into the reaction mixture. To ensure intimate mixing of the surface film with reaction mixture, tubes are shaken at an angle in the water bath. In choosing an assay temperature, the nature of the aquatic system and the aims of the investigation need to be considered.

At the termination of the assay the carriage is removed by means of the wire. With alkaline phosphatases the 4NP produced can be measured directly. Indeed, for the determination of this enzyme, the rack can be retained above the surface of the reaction mixture by means of the wire whilst the absorbance of the mixture is quickly determined, and then returned to the fluid and the reaction allowed to continue. For the determination of neutral and acid phosphatases, the rack is removed and the colour of the 4NP is developed by addition of 1 ml Na_2CO_3 (2 mol/l).

Suitable standards can be set up to derive a conversion factor for the absorbance/concentration 4NP relationship.

Fluorimetric determination

The fluorescence spectrum of 4MU in alkaline solution differs from the spectra of 4-methylumbelliferyl esters: the maximal intensity of fluorescence for MU is at 448 nm compared with emission maxima at about 380 nm for substrates (Guilbault & Hieserman 1969; Leaback 1974). The absorption characteristics of 4-methylumbelliferyl esters and 4MU are also different: the wavelengths of maximal absorption of near ultra-violet light in alkaline solution is usually about 320 nm for substrates whereas 4MU absorbs maximally at 360 nm.

Carriages to hold colonized strips are the same type as those used for colorimetric assay. The assay mixture consists of 4MUP (0.5 mmol/l) in TRIS–HCl buffer (50 mmol/l, pH 8.5). The assay system was incubated shaken at the appropriate temperature. At the end of the incubation period (the duration is dependent on the activity of the film) the carriage is removed and a sample of the incubation mixture is assayed fluorimetrically. The excitation wavelength is 360 nm and 90 degrees fluorescence is measured with an 'MSE Spectroplus' instrument using a 420 nm cut-off filter to remove fluorescent light from unhydrolysed substrate and reflected light.

When alkaline phosphatase activity is measured at pH 8.5 it is not necessary to make the solution alkaline and develop the fluorescence, but a greater fluorescence intensity may be obtained by adding 1 ml sodium glycinate buffer (200 mmol/l, pH 10.4).

The assays described here were undertaken at 25°C.

Nature of Environments Examined

For the development of the method and in subsequent applications of the method, strips were immersed at the following locations.

Location A. The bacteria bed of a small waste water treatment plant. Film formation was rapid and easily discernible after seven days exposure.

Location B. A pool in the tropical house of a botanical garden. Film formation was slow and, typically, strips were exposed for 21–28 days.

Location C. A settlement tank of a town water treatment plant. Film formation was slow. An exposure period of 15 days or longer was typical.

Reaction Kinetics for Alkaline Phosphatase in Microbial Films

Spectrophotometric Assay Technique

Rate of formation of product

Glass strips, colonized by immersion at Location A, were assayed as described. Carriages contained one, two or three strips. Fig. 2 illustrates that the rate of reaction is proportional to the quantity (area) of microbial film (i.e. the number of strips) in the reaction mixture.

Glass strips were colonized at Locations A and B. The microbial film which had

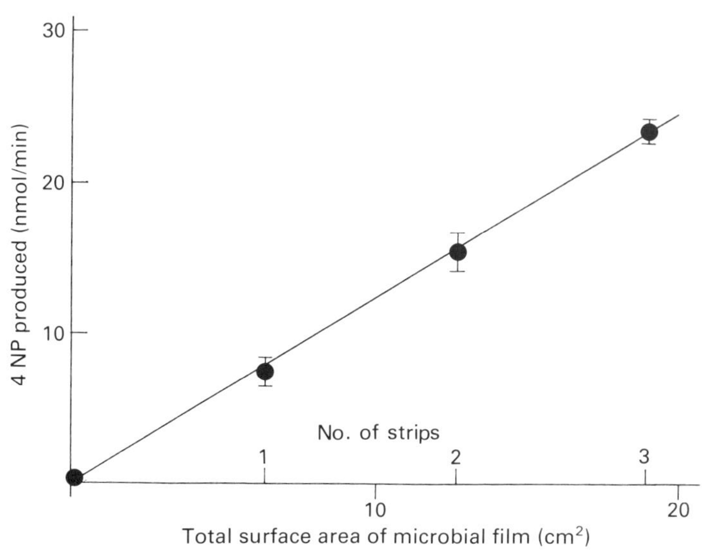

FIG. 2. Alkaline phosphatase activity of microbial films: relationship between rate of production of 4NP and area of microbial film. Glass strips colonized at Location A. Bars signify standard deviation.

developed in Location A was much heavier than that which had developed at Location B.

Fig. 3 indicates that for both sets of colonized strips, contrasting markedly in the degree of film formation, the production of 4NP was proportional to the reaction time.

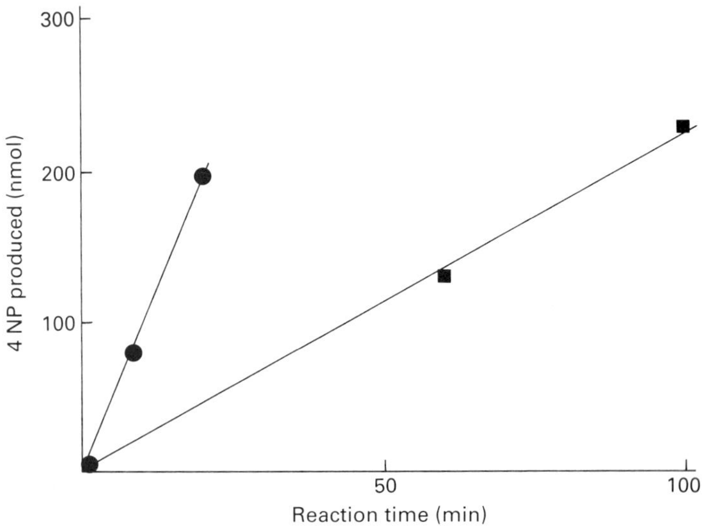

FIG. 3. Alkaline phosphatase activity of microbial films: rate of formation of 4NP of microbial films developed at different locations. Location A, film area, 6.3 cm^2 (●); Location B, film area 18.9 cm^2 (■).

Fluorimetric Assay Technique

Determinations of the effect of film quantity and of reaction time on production of 4MU were also undertaken. Glass strips had been immersed at Location C and were lightly colonized. Figs 4 and 5 show that, as with the spectrophotometric procedure, product formation is proportional to the area of film and to the reaction time.

Some Applications of the Method

Indirect determination of the progression of surface colonisation

Glass strips were immersed at Location C. Strips were removed at intervals and the alkaline phosphatase activity of the surface film determined by both the spectrophotometric and fluorimetric procedures. Fig. 6 shows that the enzyme activity per unit surface area increases with time of exposure of the strips.

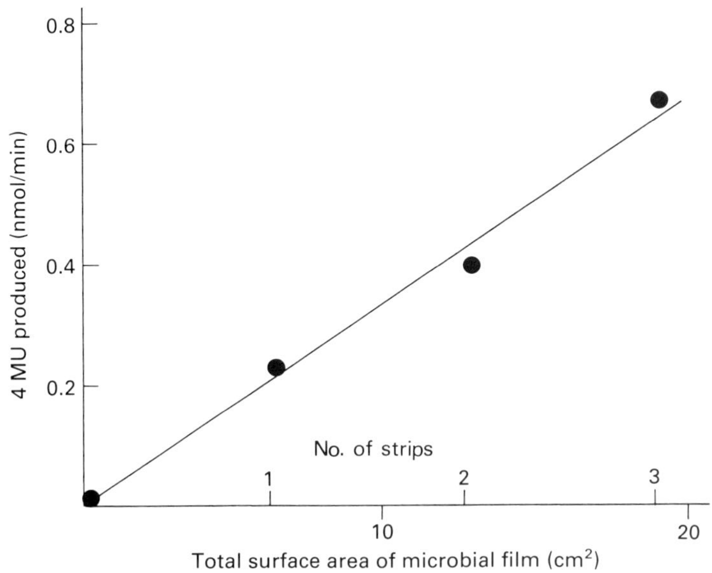

FIG. 4. Alkaline phosphatase activity of microbial films: relationship between rate of production of 4MU and area of microbial film. Glass strips colonized at Location C.

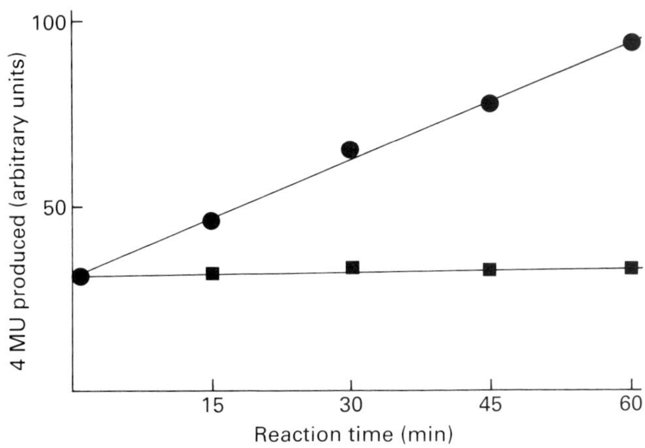

FIG. 5. Alkaline phosphatase activity of microbial films: rate of formation of 4MU by glass strips colonized at Location C. Film activity (●); background fluorescence (■).

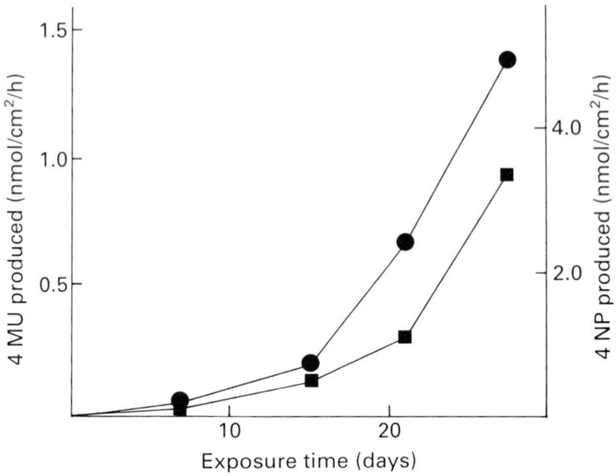

FIG. 6. Alkaline phosphatase of microbial films developing on glass strips at Location C. Activity with 4MUP substrate (●), with 4NPP substrate (■).

Microscopic examination of the strips indicated that there was a progressive increase in the cover of the strips with micro-organisms. Hence the enzyme assay procedure could afford a simple and rapid means of assessing the progression of colonization of a surface where there is a proportionality between phosphatase content and film biomass.

Determination of profiles of the effect of pH on phosphatase activity of microbial films

Glass strips were colonised by immersion at Location B. The phosphatase activity of the surface films at different pHs was then determined, using the mixed TTA buffer. The results are shown in Fig. 7.

TABLE 1. *Colonization of paint surfaces containing different biocide additives; determination by estimation of protein content and alkaline phosphatase activity of microbial films*

Paint type	Percentage values (compared with unpainted glass controls)	
	Alkaline phosphatase activity	Protein content
Glass unpainted	100	100
Copper base	25	39
Containing organic biocide	6	22
Organotin	38	28
Chlorinated rubber base	69	77

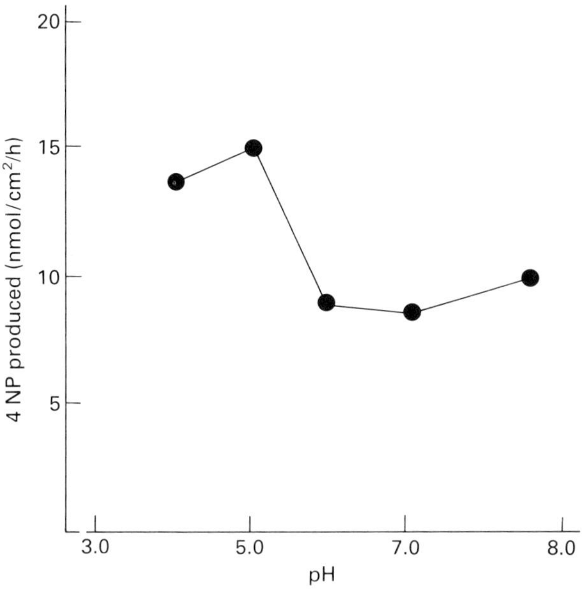

FIG. 7. Phosphatase activity at different pHs of glass strips colonized at Location B. Assay with 4NPP substrate.

The ambient pH of the water was 7.0 but the results indicate that the activity of the entire phosphatase system of the microbial film was greatest at pH 5.0. Subsequent work has shown that microbial films which have formed in different waters can differ markedly in the nature of the complement of phosphatases.

Determination of the effect of antifouling paints on microbial colonization

Glass strips were coated on both sides with paints containing different antifouling additives. The strips were colonized by immersion at Location A. Duplicate sets of the variously treated strips with their attached microbial films were then assayed for protein content (Bott & Brock 1970) and alkaline phosphatase activity using 4NPP as substrate. The protein content is indicative of the quantity of biofilm and the results in Table 1 demonstrate that the nature of the additive influences the degree to which biofilm can develop on the surface. Moreover the alkaline phosphatase assay yields activities of biofilm on the various surfaces which are in a similar order to the values for the protein content of the biofilms. Hence the results suggest that the enzyme assay procedure could afford a simple means of comparing the effectiveness of biocides in surface coatings.

Discussion

This enzymatic assay method has been developed and evaluated using glass substratum strips which had been colonized in various aquatic environments. The microbial films contain mixtures of micro-organisms and the phosphatase properties of the individual microbial types might be expected to differ. Nevertheless the examination of the reaction kinetics of the phosphatase system of such films indicates that the assay procedure can be applied with confidence. The investigations undertaken suggest that the method can be used to survey the distribution of phosphatases, and possibly other hydrolytic systems, given the availability of suitable substrates, in surface biofilms and as an indirect measure of microbial biomass.

The equipment is inexpensive and simple to construct and racks carrying substratum strips could be immersed in aquatic environments and used for a variety of monitoring purposes, e.g. for studying rates of colonization of different materials or for examining the effect of potential detrimental conditions on surface microflora.

References

Bessey, O. A., Lowry, O. H. & Brock, M. J. 1946 Method for the determination of alkaline phosphatase with five cubic millimeters of serum. *Journal of Biological Chemistry* **164**, 321–329.

Bott, T. L. & Brock, T. D. 1970 Growth and metabolism of periphytic bacteria: methodology. *Limnology and Oceanography* **15**, 333–342.

Geesey, G. G., Mutch, R. & Costerton, J. W. 1978 Sessile bacteria: an important component of the microbial population in small mountain streams. *Limnology and Oceanography* **23**, 1214–1223.

Guilbault, G. G. & Hieserman, J. 1969 Fluorimetric substrate for sulfatase and lipase. *Analytical Chemistry* **41**, 2006–2009.

Leaback, D. H. 1974 *An Introduction to the Fluorimetric Estimation of Enzyme Activities*, 2nd edn. Colnbrook: Koch-Light Laboratories Ltd.

Marshall, K. C. 1976 *Interfaces in Microbial Ecology*. Cambridge, Mass: Harvard University Press.

Monitoring Microbial Fouling in Flowing Systems using Coupons

P. D. GILBERT AND B. N. HERBERT
Shell Research Limited, Sittingbourne Research Centre, Sittingbourne, Kent ME9 8AG

Microbial Films in Aquatic Environments

A major portion of a microbial population in a flowing aquatic system may frequently be as a film colonizing a surface. This has been observed in marine (Zobell 1943, 1972; Corpe 1973) and freshwater environments (Geesey *et al.* 1977, 1978), where the sessile microorganisms may heavily outnumber their planktonic counterparts. The development of a microbial film on an initially 'clean' surface is a complex process involving several stages, each of which modifies the previous surface characteristics and influences future development of the film (Characklis & Cooksey 1983; Corpe & Winters 1980; Duddridge & Pritchard 1983). The development of a primary microbial film is thought to precede and accelerate any macrofouling. The main stages of microbial film formation may be summarized as follows:

(1) *Adsorption of an organic film.* This occurs rapidly within a few hours on exposure of a 'clean' substratum (Loeb & Neihof 1975; Kristoffersen *et al.* 1982a). The film may be composed of low molecular weight material such as amino acids, monosaccharides and fatty acids (Kristoffersen *et al.* 1982b) as well as higher molecular weight material such as glycoproteins and proteoglycans (Baier *et al.* 1974). The adsorption appears to be selective as the spectrum of compounds present on the surface can differ from those present in the bulk aqueous phase (Zsolnay & Little 1983).

(2) *Colonization by bacteria.* Bacteria may become firmly attached to the surface by means of an extracellular polysaccharide (Corpe 1970; Marshall 1973; Geesey *et al.* 1977; Zaida *et al.* 1984). Subsequent growth of these attached bacteria can lead initially to the formation of microcolonies which may later merge to form a more confluent microbial film (Jones *et al.* 1969). With time, stalked and filamentous

Industrial Microbiological Testing
0–632–01793–7

bacteria may become more numerous (Corpe 1973; Trulear & Characklis 1982). These forms may gain an advantage by extending outwards from the microbial film into the bulk water phase.

(3) *Entrapment of material.* As the microbial film develops it will entrap inorganic particles and organic detritus.

(4) *Further colonization by other microorganisms.* These may include fungi, attached and grazing protozoa, diatoms, micro-algae, etc. (Gerchakov *et al.* 1976).

(5) *Detachment.* Small portions of the microbial film may be continuously removed by shearing effects and the extent of this can depend upon flow conditions and film thickness (Trulear & Characklis 1982). Alternatively, random detachment of larger portions of the microbial film (sloughing) may be the result of nutrient depletion within the film (Howell & Atkinson 1976) or a response to gross changes in the nutrient status of the bulk water phase.

The wide occurrence of microbial films in flowing systems and the numerical predominance of the microorganisms within them would suggest that this mode of existence confers some advantage over planktonic microorganisms.

In nutrient-poor waters, the accumulation of an organic film at the substratum–water interface may provide nutrient concentrations and growth rates unattainable in the bulk water phase. Once established on a surface, continued accumulation of nutrients from the flowing bulk water phase may be a means of avoiding nutrient depletion that could occur within a unit volume of the bulk water phase. Furthermore in a mature microbial film the close juxtaposition of bacteria and micro-algae may facilitate any trophic relationships between these two groups. In the bulk water phase the benefit of similar relationships may be diminished due to the greater distances over which they have to act. Contact of bacteria with a substratum has been shown to influence metabolic activity. For example, a substratum or its commonly associated surfactants can cause an increase in metabolic activity (Kjelleberg *et al.* 1983; Humphrey & Marshall 1984). Naturally occurring substrata were found to support larger more frequently dividing bacteria that were responsible for the majority of the carbon source uptake compared to planktonic bacteria (Pedros-Alio & Brock 1983).

Apart from its possible nutritional advantage a microbial film may also provide microenvironments where conditions are more favourable for microbial growth/activity than those prevailing in the bulk water phase. This can largely be attributed to the extracellular polysaccharide matrix in which the bacteria within the microbial film are commonly embedded. This can give rise to gradients between the bulk water phase and the substratum, such as pH and dissolved oxygen, which may favour different members of the microbial film community. For example oxygen depletion in the lower levels of the film can create anoxic regions close to the substratum which permit the activity of anaerobic species such as sulphate-reducing bacteria.

Importance of the Microbial Film

The presence of microbial films is not restricted to naturally occurring substrata in the environment. Their development on man made objects with potential detrimental effects is one of the main reasons for interest in microbial films. Microbial films on the external surfaces of marine structures such as pilings, offshore oil installations and ships' hulls are thought to precede and accelerate heavier macrofouling. This may lead to increased stresses in the structure resulting from an increase in weight and drag, as well as a potential increase in corrosion risk underneath the fouling. These problems are mainly related to the development of macrofouling and lie beyond the scope of this article.

Microbial films can also be found on the internal surface of pipework carrying flowing water such as heat exchangers, crude oil transport systems, and water injection systems used in secondary oil recovery. In these cases possible detrimental effects of the microbial film may include: (1) An increase in fluid frictional resistance, thereby raising the energy needed to maintain a given flow rate. (2) Impairment of heat transfer. (3) Enhancement of corrosion due to the physical presence and metabolic activity of the microbial film. (4) Blockage of filters either from development of a microbial film on the filter matrix or following sloughing of the microbial film from other areas of the system. (5) Reduction in the value of the product being transported in the pipeline; for example, hydrogen sulphide production by a microbial film can lead to souring of a gas or crude oil while in transit.

The importance of microbial films associated with the corrosion of condensers in the power industry is well documented (Tarbuck & Wyborn 1980; Sakaguchi 1980; Bibb & Hartmann 1984). Sulphate-reducing bacteria are often implicated in steel corrosion under a microbial film (Gaylarde & Johnston 1980; Sanders & Maxwell 1983; Hamilton & Sanders 1984).

In such cases the activities of other microbial species in the microbial film may result in oxygen depletion and supply carbon/energy sources required for the activity of sulphate-reducing bacteria. Other microbial groups such as iron-reducing bacteria (Obuekwe *et al.* 1981) and micro-algae (Terry & Edyvean 1984) have also been implicated in enhanced corrosion under microbial films.

A growing awareness of the possible additional detrimental effects of microbial films as described above has led to the requirement for some means of *in situ* sampling which can enable subsequent examination of a portion of the microbial film. This is needed for both general monitoring and to assess the effectiveness of possible control measures. Traditionally, measurement of microbial numbers in industrial systems has largely been restricted to the planktonic population. It is on the basis of these measurements that the performance of biocide treatments has been assessed. However, monitoring of the microbial film has the following advantages:

(1) The microbial film may represent the majority of microorganisms in the system in terms of numbers or biomass (Pedros-Alio & Brock 1983; Geesey *et al.* 1978). Furthermore, the species which predominate in the bulk water phase may be different from those present in the microbial film.

(2) The physiological activity, for example growth rate, of bacteria in the microbial film may differ from that of similar bacteria in the bulk water phase.

(3) The susceptibility of microorganisms in the microbial film to antimicrobial agents, such as biocides, may differ from equivalent microorganisms dispersed in the bulk water phase. In the case of microbial films which developed in some oilfield operations, higher concentrations of biocide were required to show activity against bacteria present in the microbial film compared to those in the bulk water phase (Ruseska *et al.* 1982; Costerton & Lashen 1984). This difference was largely attributable to protective effects of the extracellular polysaccharide matrix surrounding bacteria in the microbial film which may have hindered penetration of biocide to the cell surface.

The renewed interest in microbial films has stimulated the development of a number of devices that can be used to monitor their development. Some are only suitable for use in the laboratory and can be used to investigate microbial film development and the formulation of control measures. Others have been specifically designed for monitoring in industrial situations and permit withdrawal of a portion of the microbial film from enclosed systems, such as pipelines, for subsequent examination.

Sampling Devices

There are several commercially available sampling devices for monitoring microbial film formation. These fall into two main types, namely directly implanted and side stream devices. References to suppliers are given at the end of this chapter. With the exception of the Cosasco glass slide holders, all use retrievable circular metal pieces (coupons) as the surface for microbial film development and subsequent examination. The coupons are mounted in holding assemblies which are then inserted into the pipework in such a way that the coupons are, as closely as possible, flush mounted with the internal wall surface of the pipe. As flow characteristics have a major influence on microbial film development, this flush mounting is essential to ensure that the coupons experience a similar flow regime to the rest of the pipe surface. This will, in turn, increase the probability of a microbial film, representative of the microbial film present on most of the pipe surface, developing on the coupons. The entire assembly can be withdrawn at intervals and the coupon(s) removed for examination.

Directly Implanted Devices

Most of these assemblies are designed to be compatible with existing high pressure fittings, e.g. the Cosasco 2 in. access fitting assembly. The sampling devices can be serviced at the pressure within the system, thereby avoiding a partial shut down and depressurization. Figs. 1 and 2 show two typical sampling devices. The exposed face of the metal coupons has a surface area of 0.5 cm^2 and the entire coupon can be manufactured of different materials such that it can be identical to the material of the monitored system. Coupons are removed from the holding assembly from the back, sometimes by means of a special tool, and this avoids any direct contact with the microbial film present on the coupon face.

A glass slide holding assembly (Fig. 3) provides an alternative to the use of metal coupons. However, this suffers from the disadvantage that it is not flush-mounted.

Side Stream Devices

These are mounted in parallel to the main system and take a proportion of its flow. They can be isolated from the main system prior to sampling. This avoids the need for the special fittings and tools needed to gain access under pressure. Fig. 4 gives details of a typical side stream sampler (Robbins device). Metal coupons identical to those used in the directly implanted assemblies are used, but the device holds a greater number and they are mounted individually in separate holders. The side stream sampler can give greater flexibility than directly implanted assemblies. The larger number of coupons enables samples to be withdrawn more frequently which allows closer monitoring and the use of several coupons to check the uniformity of the sampled microbial film. In addition side stream samplers can be isolated from the main flow and subsequently connected to experimental flow loops for treatment (e.g. biocide) testing. In this way the sampled microbial film can be exposed to any treatment testing without being subject to disturbances which may occur in the main system, or to those caused by removing the coupon from the system.

Originally the Robbins device was only available constructed from Admiralty brass. Doubts that the microbial films formed were representative of those which developed on the actual pipe surface, which may be constructed of a different material, have led to this device losing favour. However, it can now be obtained in different materials, e.g. mild steel.

Our own needs in sampling from small bore (3 in. dia) pipe in water injection systems on offshore oil production platforms has led to the development of a different type of sampling device. In this case the metal coupons are mounted radially around the circumference of a holding assembly of identical internal diameter to that of the pipe. This enables flush mounting to be achieved. The coupon holding assembly was constructed from a modified carrier ring assembly

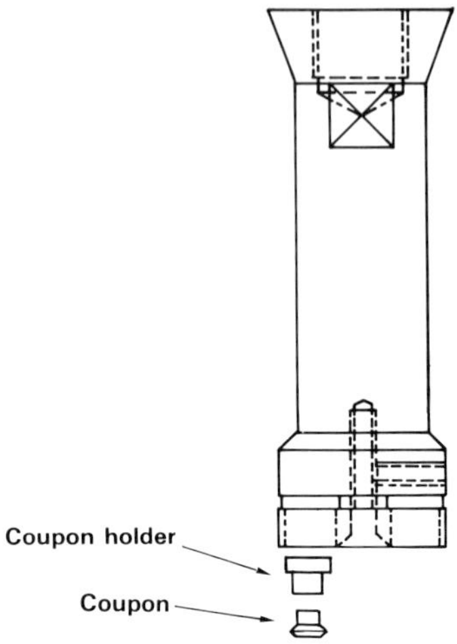

FIG. 1 (a). Caproco biofilm probe.

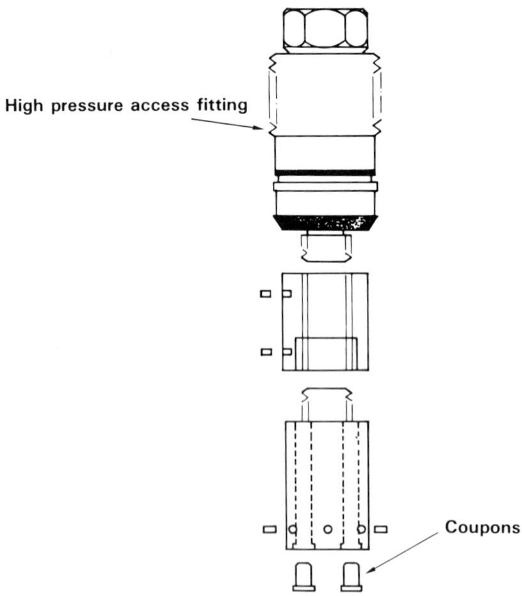

FIG. 1 (b). Petrolite bioprobe with Cosasco solid plug.

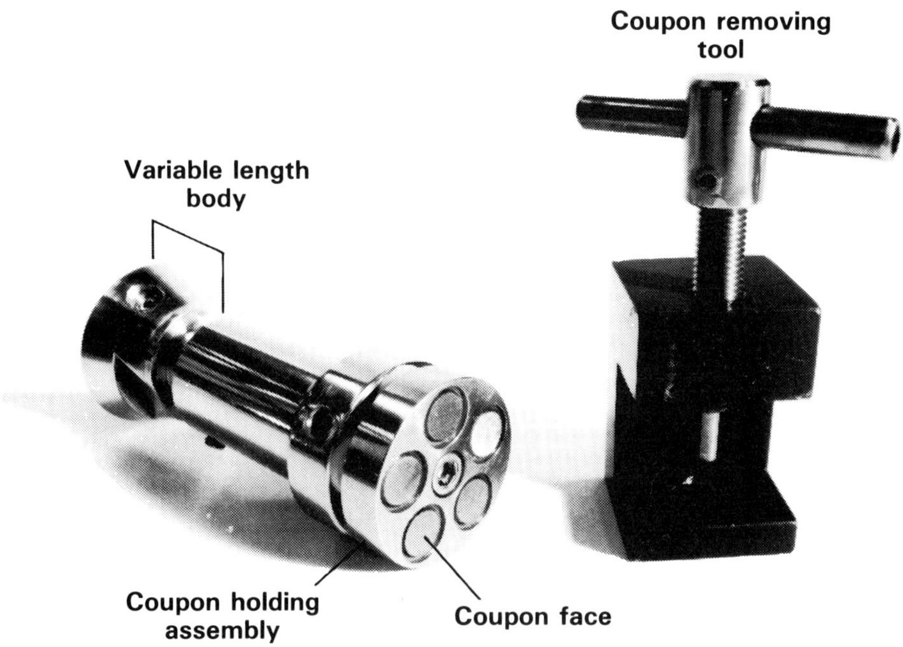

(a) Caproco biofilm probe with coupon removing tool.

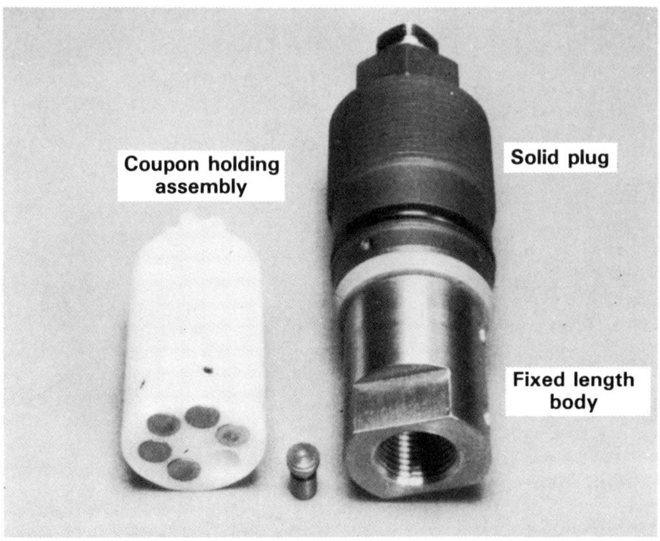

(b) Petrolite bioprobe with Cosasco solid plug.

FIG. 2. Photographs of two commercially available probes.

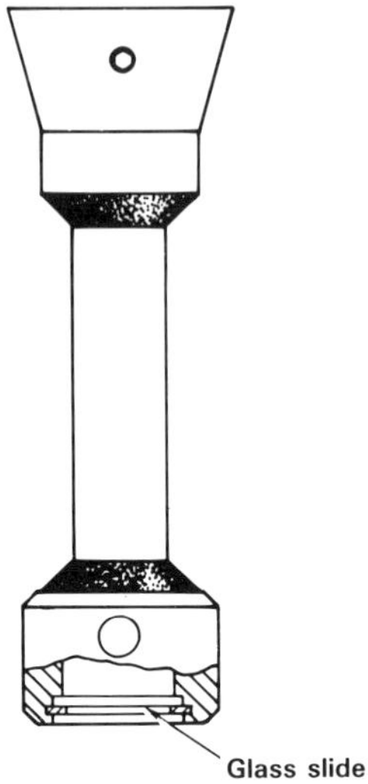

FIG. 3. Cosasco glass slide assembly.

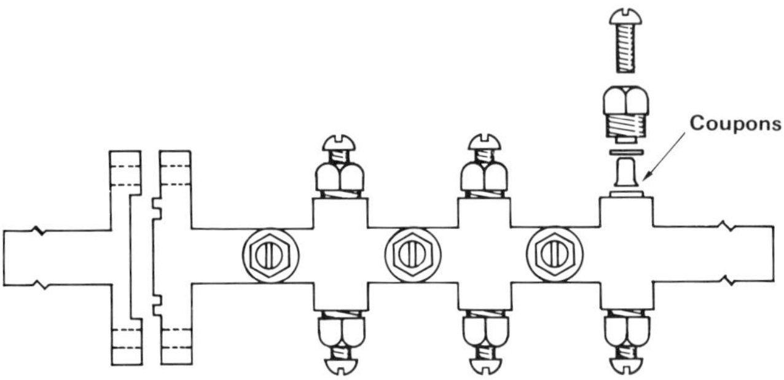

FIG. 4. Robbins device.

which, in turn, forms the removable portion of an orifice plate holder. This system has the disadvantage that the coupons can not be withdrawn under pressure and hence partial shut down and depressurization is required prior to sampling.

Laboratory-based Fouling Rigs

Laboratory-based fouling rigs allow a more detailed and fundamental study of the microbial film phenomenon. They permit control of important parameters such as temperature, flow velocity, nutrient status, material type, etc., enabling the effects of these on microbial film development/activity to be determined. In addition, they can be used to establish microbial films which can subsequently be used to investigate control strategies such as biocide treatment (Ruseska *et al.* 1982).

Apart from apparatus specifically designed to study microbial attachment such as the radial flow chamber (Fowler & McKay 1980), most laboratory fouling rigs are of the recirculating 'feed and bleed' type. As the name suggests, the majority of the liquid volume of the fouling portion of the rig is circulated with the continuous addition of a small volume of feed (fresh medium) and removal of an equal volume of bleed (spent medium and culture). An alternative to a continuous feed is the semicontinuous addition of a larger volume of batch feed. This may be useful where the feed is an environmental sample which cannot be supplied on a continuous basis. Whatever the nature of the feed, a recirculating system overcomes the problem of the impractically immense volume of water that would be required to run a 'once through' system at high flow velocities.

Equipment of the continuous feed and bleed type is commonly divided into two, a first stage where bacteria or other microorganisms are cultivated in continuous culture, prior to being fed into a second stage fouling reactor into which the substrata to be fouled are placed. Some of the advantages of using continuous culture in the first stage can be summarized as follows:

(1) For a given set of conditions in the first stage the inoculum provided for the second stage should be uniform and in a constant physiological state. This should minimize any experimental variation in the second stage resulting from inoculum quality.

(2) Fresh substrata for fouling can be introduced into the second stage throughout the experiment. These should theoretically experience the same 'fouling' environment independent of the time at which they are introduced into the system.

(3) Conditions in the first stage can be varied to study the effect of inoculum condition on fouling. For example, growth rate and physiological responses to different types of nutrient limitation can be determined by varying the dilution rate and medium composition in the first stage. These parameters can also be used to alter the relative proportions of different members of a mixed population within the first stage culture. This may be particularly important in consortia of microorganisms, e.g. where general heterotrophic activity is required to support subsequent microbial sulphate reduction.

In many cases the original inoculum used to start the continuous culture is taken from the environment where the fouling occurs. Some workers include a preselection for tactophilic organisms during the first stage by including a large surface area, e.g. glass beads, rocks, etc., in the continuous culture vessel. The effluent from this stage often contains larger numbers of bacteria and higher nutrient concentrations than those that would be found in the natural fouling environment. This can be overcome by dilution of the effluent with a suitable medium prior to feed into the second stage.

Substrata to be subject to fouling are placed in the second stage of the equipment. This generally consists of a long tube or sections of tube connected to a pump capable of circulating the contents at a high flow velocity which may be controlled by varying the speed of the pump or by including a regulated by-pass circuit. Bleed from the second stage may be independently matched to the feed from the first stage (i.e. input from the continuous culture and dilution medium). Alternatively an external weir connected to the dilution vessel will automatically maintain a constant volume. The high flow velocities may lead to heat being generated and hence a cooling circuit may have to be included.

Retrievable coupons such as those used in commercial fittings described previously can provide the substrata in the second stage reactor. These have the advantages that they are directly comparable to the field situation and their manageable size facilitates their inclusion into a variety of tests. Disadvantages arise from the fact that microbial film formation on implanted coupons is seldom uniform and therefore several have to be withdrawn from the system to obtain meaningful results. Also there always remains some doubt as to whether the coupons are mounted sufficiently flush with the tube wall to experience the same flow pattern as an unaltered pipe. A further problem can arise during biocide testing with coupons *in situ* in the second stage. In some cases the material of the coupons can differ from that of the main reactor tube and their surface area may represent only a small proportion of the second stage. In such cases the microbial film developed on the coupons may not be representative of the bulk of the microbial film present in the system. The effect of any biocide treatment applies to the total microbial film contained in the entire second stage. Therefore any assessment of biocide demand, derived from retrieved coupons, may reflect the biocide requirement for the system as a whole which may differ from the biocide demand of the coupons.

The use of continuous removable sections of pipe instead of retrievable coupons in the second stage can overcome some of the disadvantages of the latter (Tanner *et al.* 1985). Microbial film formation on pipe sections is often more uniform than that produced on coupons and therefore provides a better starting material for further tests. The uncertainty over flush mounting of coupons is avoided and most of the microbial film developed in the second stage is present on the pipe sections under test which increases the validity of the results of *in situ* biocide testing.

Fig. 5 shows a diagram of a typical fouling rig with alternative pipe section (A) or coupon (B) second stages. Fig. 6 shows a two-stage retrievable coupon system.

Methods for Quantification of the Microbial Film

Once metal coupons have been retrieved from the system, examination and quantification of microbial film development on the exposed face can be undertaken by a variety of methods. These include both direct and indirect methods as follows:

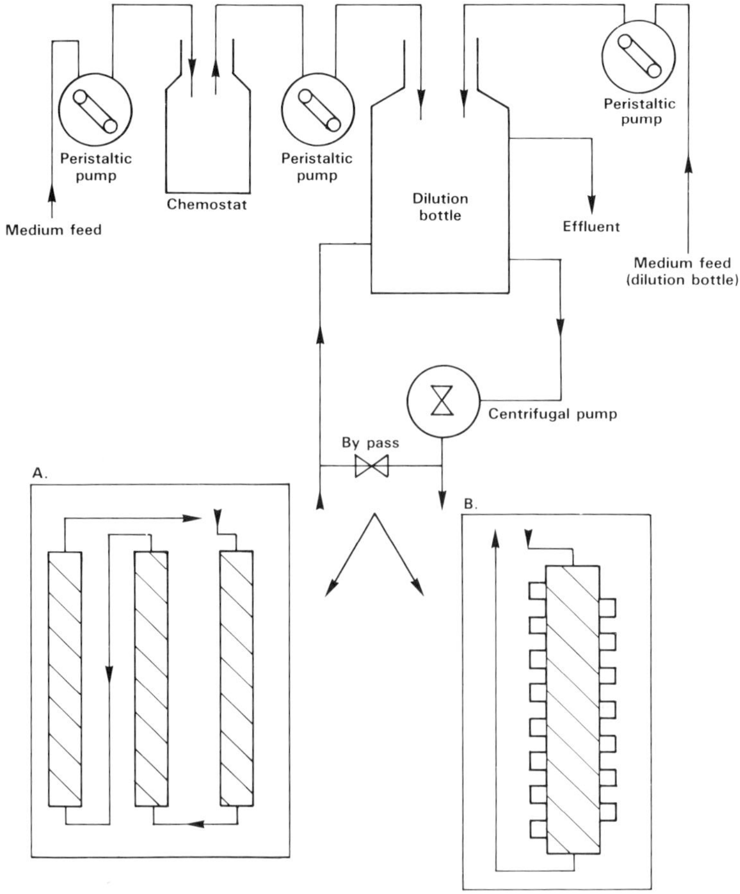

FIG. 5. Typical two-stage laboratory fouling rig showing alternative second stage reactors. (A) Second stage comprised of removable segments. (B) Tube containing retrievable coupons, i.e. Robbins type device.

Direct Examination

This can be undertaken by scanning electron microscopy and incident light microscopy.

Scanning electron microscopy

Samples may be prepared as follows:

(1) Fixation in a 2.5% glutaraldehyde solution (1 h) prepared in water of similar composition to that in the system being sampled, e.g. sea water for marine systems, fresh water for freshwater systems.

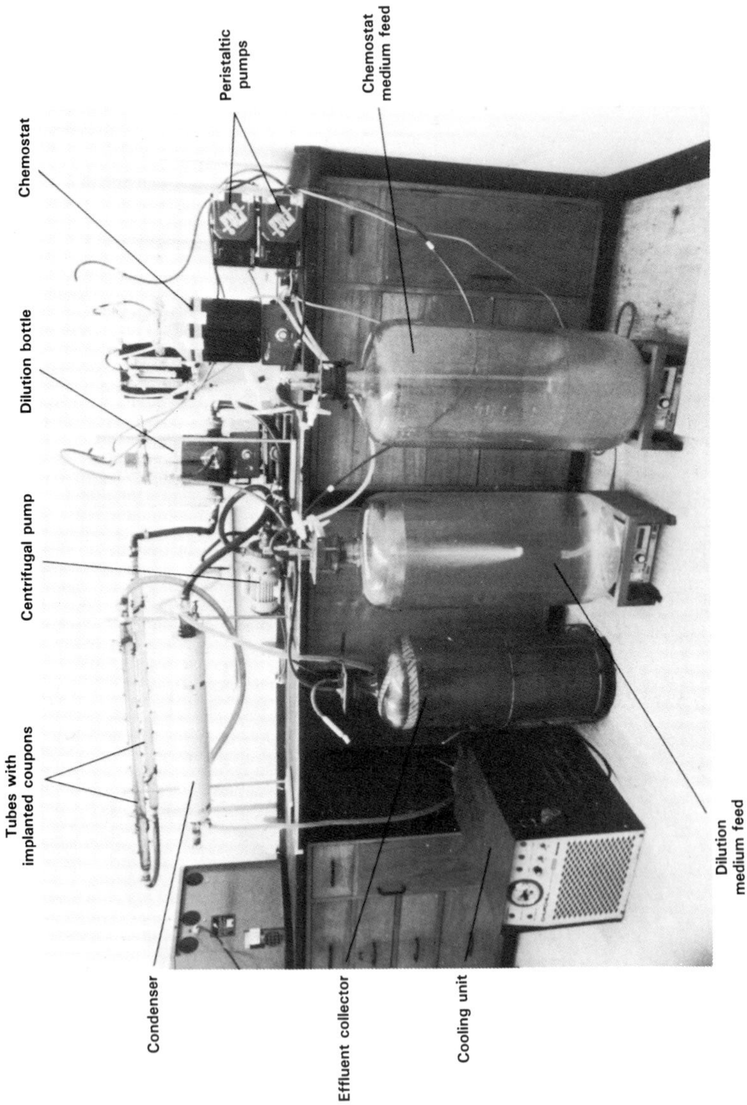

FIG. 6. Typical two-stage laboratory-based fouling rig.

(2) Desalination in a graded series (20–100%) of distilled water (20 min each dilution) for samples originating from high salinity waters, e.g. sea water, oilfield brines.

(3) Dehydration in a graded ethanol series (20–100%; 20 min each dilution) followed by air drying. Improved preservation of surface detail may be achieved by following the ethanol dehydration immediately with further dehydration in a graded Freon 13 series (30–100%) followed by critical point drying in Freon 13. Alternatively, following fixation and desalination (if required) the sample can be dehydrated by freeze-drying.

(4) Coating: samples should be gold coated prior to examination (e.g. Polaron sputter coater, 20 mA for 3 min). Following coating, samples are ready for examination.

An alternative method of sample preparation and examination is the use of cryo-scanning electron microscopy. In this method the sample is treated by cooling in liquid nitrogen slush followed by low temperature (−160°C) gold coating and examination (Beckett *et al.* 1982). This method may provide minimum distortion of surface detail but does require specialized low temperature coating and stage mounting equipment.

Light microscopy

The nature of the coupons prevent *in situ* observation of bacteria in the microbial film by conventional staining and transmission light microscopy techniques. Epifluorescence (incident light) microscopy provides an alternative method of *in situ* observation that requires less specialized and cheaper equipment than that necessary for scanning electron microscopy but will not, of course, provide as great a magnification.

Bacteria are stained with a fluorescent dye and observed under an incident light source and filter combination selected to produce fluorescence of the stained bacteria. A variety of fluorescent stains are available but acridine orange is widely and successfully used (Daley & Hobbie 1975; Hobbie *et al.* 1977; Jones & Simon 1975). Typically, bacteria may be stained for 2–5 min in dye concentrations of 5–10 mg/l^{-1}.

Refinements of the epifluorescence technique include the use of fluorochromes such as 2-(p-iodophenyl)-3-(p-nitrophenyl)-5-phenyl tetrazolium chloride (INT) and fluorescein diacetate which are stains specifically formulated to identify actively metabolizing cells (Tabor & Niehof 1982; Quinn 1984). Additionally, immunofluorescence techniques provide a means of identifying specific microbial groups within the film (Smith 1982; Zambon *et al.* 1984).

Indirect Methods

Indirect methods of microbial film quantification include activity measurements, estimation of biomass, and enumeration of viable bacteria by culture techniques.

Activity measurements

Radiorespirometric techniques provide a sensitive and rapid means of determining microbial activity within the microbial film. They also have the advantage of being performed with the microbial film *in situ* on the retrieved coupon. This entails minimum disturbance of the microbial film and therefore reduces any deleterious effects of sample handing such as loss of viability/activity, drying, etc. Activity measurements are useful in that some methods can determine the activity of specific groups of bacteria, e.g. sùlphate-reducing bacteria, within the microbial film.

Total heterotrophic activity. This can be estimated by respiration of [U-^{14}C]D-glucose to [^{14}C]CO_2. In most cases the concentration of endogenous glucose in the entire sample (coupon, associated microbial film and an aliquot of system water) will be negligible and therefore unlabelled 'cold' glucose will have to be added to the assay. This means that the assay will determine the potential of the microbial film for a glucose respiration rather than a true *in situ* rate. The latter would be difficult to determine in any case as the assay requires the coupon to be removed from the flowing system. This means that the microbial film is removed from a continuous supply of nutrient present in the aqueous phase of the flowing system, and separated from the rheological characteristics of the system which determine nutrient transport to the microbial film. Despite these limitations, determination of the potential heterotrophic activity of the microbial film is still a useful parameter in microbial film quantification.

Typically the assay may be performed as follows:

(1) A coupon together with 2.5 ml of water is taken from the system and placed in the assay tube (Fig. 7). The bung complete with wick holder and wick is then crimp-capped into place.

(2) Stock solution of labelled [^{14}C]D-glucose and 'cold' glucose are added to give final concentrations of 1 μCi and 50 μM respectively. Care should be taken not to contaminate the wick with labelled glucose solution.

(3) Assay tubes are incubated for 6 h in a reciprocating water bath held at a temperature close to that of the system under examination.

(4) Trapping of evolved CO_2 is initiated by adding 0.8 ml 1 M KOH (CO_2 trapped as HCO_3^-) to the wick and wick cup holder followed by 1.0 ml 6 M HCl added to the solution containing the coupon to promote evolution of CO_2. Trapping may be complete after 6 h incubation of the assay tubes in the reciprocating water bath with the temperature adjusted to 30–35°C.

(5) Following trapping, wicks are removed from the wick holder and placed in scintillant (e.g. Instagel; Packard Instruments) together with any residual trapping agent and a distilled water washing from the wick holder. The trapped [^{14}C]HCO_3^- is then determined by liquid scintillation counting.

Sulphate reduction (sulphate-reducing bacteria). The activity of these bacteria is of

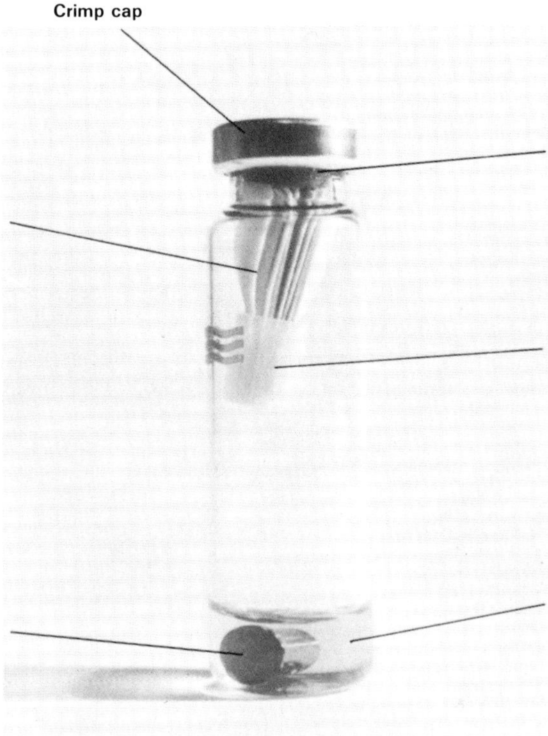

FIG. 7. Typical assay tube for radiorespirometric studies.

particular interest in microbial films because of the association of the bacteria with metal corrosion (Tiller 1983; Cragnolino & Tuovinen 1984). The radiorespirometric method of activity measurement (Rosser & Hamilton 1983; Hardy & Syrett 1983) relies upon reduction of $[^{35}S]SO_4^{2-}$ to $[^{35}S]S^{2-}$ which is subsequently trapped and determined by liquid scintillation counting. As with the assay to determine total heterotrophic activity, it is not strictly correct to claim a true *in situ* rate of sulphate-reduction once the coupon has been removed from the system. An approximation can be made by incubating the coupon in source water from the system. But even in this case, nutrient limitation, and hence a reduction in activity, may occur which is not experienced in the flowing system. For calculation of a rate of sulphate reduction the endogenous sulphate concentration must be determined. The assay is similar to that used for $[^{14}C]$D-glucose respiration with the following exceptions:

(1) Immediately after placing the coupon and source water in the assay tube and capping, the tube is degassed under vacuum and the head space flushed three times with oxygen-free nitrogen (the third change of gas remaining in the tube to

provide an anaerobic headspace). More stringent anaerobic conditions can be obtained by the addition of a $Na_2S_2O_4$ solution (final concentration 2 mM). A stock solution should be prepared freshly in NaOH solution to give a neutral pH.

(2) The amount of labelled $[^{35}S]Na_2SO_4$ added to each tube will depend upon the arbitrary judgment of the expected activity of the sulphate-reducing bacteria within the microbial film and knowledge of the sulphate concentration in the source water used in the assay. For example, a coupon supporting a light microbial film incubated in source water with a high sulphate concentration (e.g. marine systems) may require the addition of more label to produce a meaningful count than a heavily fouled coupon incubated in source water with a low sulphate concentration. Addition of between 1 and 5 μCi should accommodate most cases.

(3) Zinc acetate (1.0 M) is used as a trapping agent but this should be prepared and used as an anaerobic solution, likewise the 6 M HCl. Trapping time should be increased to 16 h.

The potential for sulphate-reduction can be determined by adding a carbon/energy source such as lactate (0.1% w/v) or dextrose (0.1% w/v) and yeast extract (0.01% w/v) to the assay medium. With both radio-respirometric assays an increase in pressure may occur in the assay tubes resulting from hydrogen evolution from the metal coupon under the acidic conditions required for trapping.

Indirect Measurements of Biomass

Direct measurement of biomass within the microbial film may prove difficult because of the small mass of microbial film compared to that of the coupon and the entrapment of inorganic particles such as corrosion products or inorganic precipitates (scale). Microscopical examination can give an estimate of microbial film biomass. However, there are several techniques applicable to indirect determination of biomass. These include measurement of ATP (Holm-Hansen & Booth 1966), muramic acid (King & White 1977), and lipopolysaccharide (Watson *et al.* 1977). All these methods require prior removal of the microbial film from the coupon.

Culture Techniques

Conventional microbiological culture techniques provide a means of estimating the number of viable bacteria present in the microbial film and therefore give an indication of the extent of fouling. They can be extended to include diagnostic media which can be used to identify the presence of certain bacterial types. They have the advantage that they are easy to perform and interpret, and do not require specialized equipment. However, they also suffer from several disadvantages, namely:

(1) The microbial film has to be removed from the coupon and homogenized to

disperse clusters of cells prior to counting for viable cells. This is difficult to achieve without loss of viability.

(2) Bacteria which predominate numerically within the microbial film may not be the most important in terms of activity.

(3) Disruption of the microbial film may destroy intricate nutritional relationships between different bacterial groups which are not reproducible in defined culture media.

Several methods for microbial film removal have been recommended. Scraping with a sterile instrument such as a scapel blade may remove the film which should then be homogenized to disperse cells prior to counting. Mild sonication gives a combined method of removal and dispersion but over application can cause a loss in viability.

Enzymic treatment of the microbial film with polysaccharases may weaken the intracellular polysaccharide matrix and aid removal/dispersion in conjunction with sonication. Despite the sophistication of the various techniques, recovery of viable cells in relation to total cell numbers is reported to be low (Duddridge, personal communication). In addition, many of the bacteria in the microbial film may be lodged in crevices and therefore virtually impossible to remove by scraping or mild sonication.

After removal from the coupons the numbers of viable suspended bacteria can be determined by plating out onto suitable solid media or by serial dilution in broths using standard techniques (American Petroleum Institute 1965).

In addition to the described procedures, microbial films formed in laboratory rigs can be quantified by weight determination, microscopic examination of film thickness, and increase in system pressure (Bryers & Characklis 1980).

Conclusions

The recognition that attached microorganisms have an important role in flowing systems has led to an increased interest in the application of the types of monitoring systems that have been described in this review. The various limitations of the currently available systems are beginning to be recognized, particularly with respect to their use on offshore oil-producing platforms. It is hoped that future work will be aimed at improving reproducibility of microbial film development in laboratory rigs and in providing simpler methods for assessing microbial films on coupons retrieved from the field which may have to be carried out by unskilled personnel.

Measurements of microbial films should not be carried out without complementary assessment of the planktonic microbial population. It is only by taking both measurements into account that an overall understanding of the pipeline environment can be gained.

It should be recognized that the microbial films obtained in the laboratory rig can be very different from those that develop in the field. In the field situation, e.g. a pipeline in an oilfield operation, the microbial challenge will be uneven and the flow rate may vary and be subject to stoppage for operational reasons. In addition, the microbial film will be developing in an environment where other particulate material may be suspended in the water and will be deposited on the substrata together with the microbial film. Such materials could be paraffin waxes, sand and scaling chemicals (e.g. calcium or barium sulphate).

Acknowledgments

The authors would like to acknowledge the following suppliers of sampling devices for the provision of information, diagrams and photographs describing their equipment: Petrolite Ltd, Birchill Road, Kirby Industrial Estate, Liverpool L33 7TD; Caproco Corrosion Prevention Ltd, 56–58 Watford Way, London NW7 4RS; Rohrback Cosasco Systems Ltd, Rohr. Co. House, Unit A1, Worton Grange, Imperial Way, Reading, Berks RG2 0TD.

References

American Petroleum Institute (API) 1965 *Recommended Practice for Biological Analysis of Subsurface Injection Waters*, 2nd edn. API RP No. 38. Dallas: Division of Production.

Baier, R. E., Goupil, D. W., Perlmutter, S. & King, R. 1974 Dominant chemical composition of sea-surface films, natural slicks and foams. *Journal de Recherches Atmospheriques* **8**, 571–600.

Beckett, A., Porter, R. & Read, N. D. 1982 Low temperature scanning electron microscopy of fungal material. *Journal of Microscopy* **125**, 193–199.

Bibb, M. & Harmann, K. W. 1984 Bacterial corrosion. *Corrosion and Coatings, South Africa* **11**, 12–19 and 29.

Bryers, J. D. & Characklis, W. G. 1980 Measurement of primary biofilm formation. *Proceedings of the Symposium on Condenser Biofouling and Control*, eds Garey, J. F. *et al.*, pp. 169–183. Michigan: Ann Arbor Science.

Characklis, W. G. & Cooksey, K. E. 1983 Biofilms and microbial fouling. *Advances in Applied Microbiology* **29**, 93–138.

Corpe, W. A. 1970 An acid polysaccharide produced by a primary film forming marine bacterium. *Developments in Industrial Microbiology* **11**, 402–412.

Corpe, W. A. 1973 Microfouling: the role of primary film forming marine bacteria. *Proceedings of the 3rd International Congress on Marine Corrosion and Fouling*, eds Acher, R. F. *et al.*, pp. 598–609. Evanston, Ill: Northwestern University Press.

Corpe, W. A. & Winters, H. 1980 The biology of microfouling of solid surfaces, with special reference to power plant heat exchanges. *Proceedings of the Symposium on Condenser Biofouling and Control*, eds Garey, J. F. *et al.*, pp. 29–42. Michigan: Ann Arbor Science.

Costerton, J. W. & Lashen, E. S. 1984 Influence of biofilm on efficacy of biocides on corrosion-causing bacteria. *Materials Performance* Feb., 13–17.

Cragnolino, G. & Tuovinen, O. H. 1984 The role of sulphate-reducing and sulphur-oxidizing bacteria in the localized corrosion of iron-base alloys. A review. *International Biodeterioration* **20**, 9–26.

DALEY, R. J. & HOBBIE, J. E. 1975 Direct counts of aquatic bacteria by a modified epifluorescence technique. *Limnology and Oceanography* **20**, 875–882.

DUDDRIDGE, J. E. & PRITCHARD, A. M. 1983 Factors affecting adhesion of bacteria to surfaces. *Microbial Corrosion. Proceedings of the Conference on Microbial Corrosion, 8–10 March 1983, NPL, Teddington*, pp. 28–36. London: The Metals Society.

FOWLER, H. W. & MCKAY, A. J. 1980 The measurement of microbial adhesion. *Microbial Adhesion to surfaces*, eds Berkeley, R. C. W. *et al.*, pp. 143–161. Chichester: Ellis Horwood.

GAYLARDE, C. C. & JOHNSTON, J. M. 1980 The importance of microbial adhesion in anaerobic metal corrosion. *Microbial Adhesion to Surfaces*, eds Berkeley, R. C. W. *et al.*, pp. 511–513. Chichester: Ellis Horwood.

GERCHAKOV, S. M., MARSZALK, O. S., ROTH, F. J. & UDEY, L. R. 1976 Succession of periphytic microorganisms on metal and glass surfaces in natural seawater. *Proceedings of the 4th International Congress on Marine Corrosion and Fouling, Antibes-Juan-Les-Pins*, pp. 205–211.

GEESEY, G. G., MUTCH, R. & COSTERTON, J. W. 1978 Sessile bacteria: an important component of the microbial population in small mountain streams. *Limnology and Oceanography* **23**, 1214–1223.

GEESEY, G. G., RICHARDSON, W. T., YEOMANS, H. G., IRVIN, R. T. & COSTERTON, J. W. 1977 Microscopic examination of natural sessile bacterial populations from an alpine stream. *Canadian Journal of Microbiology* **23**, 1733–1736.

HAMILTON, W. A. & SANDERS, P. F. 1984 Sulphate-reducing bacteria and anaerobic corrosion. *Corrosion and Marine Growth in Offshore Structures*, eds Lewis, J. R. & Mercer, A. D., pp. 23–30. Chichester: Ellis Horwood.

HARDY, J. A. & SYRETT, K. R. 1983 A radiorespirometric method for evaluating inhibitors of sulphate-reducing bacteria. *European Journal of Applied Microbiology and Biotechnology* **17**, 49–52.

HOBBIE, J. E., DALEY, R. J. & JASPER, S. 1977 Use of nucleopore filters for counting bacteria by fluorescence microscopy. *Applied and Environmental Microbiology* **33**, 1225–1228.

HOLM-HANSEN, O. & BOOTH, C. R. 1966 The measurement of adenosine triphosphate in the ocean and its ecological significance. *Limnology and Oceanography* **11**, 510–519.

HOWELL, J. A. & ATKINSON, B. 1976 Sloughing of microbial film in trickling filters. *Water Research* **10**, 307–315.

HUMPHREY, B. A. & MARSHALL, K. C. 1984 The triggering effect of surfaces and surfactants on heat output, oxygen consumption and size reduction of a starving marine vibrio. *Archives of Microbiology* **140**, 166–170.

JONES, H. C., ROTH, I. L. & SANDERS, W. M. 1969 Electron microscopic study of a slime layer. *Journal of Bacteriology* **99**, 316–325.

JONES, J. G. & SIMON, B. M. 1975 An investigation of errors in direct counts of aquatic bacteria by epifluorescence microscopy with reference to a new method of dyeing membrane filters. *Journal of Applied Bacteriology* **39**, 317–329.

KING, J. D. & WHITE, D. C. 1977 Muramic acid as a measure of microbial biomass in estuarine and marine sample. *Applied and Environmental Microbiology* **33**, 777–783.

KJELLEBERG, S., HUMPHREY, B. A. & MARSHALL, K. C. 1983 Initial phases of starvation and activity of bacteria at surfaces. *Applied and Environmental Microbiology* **46**, 978–984.

KRISTOFFERSEN, A., ROLLA, G., SKJORLAND, K., GLANTZ, P. O. & IVARSSON, B. 1982a Evidence for the formation of organic films on metal surfaces in seawater. *Journal of Colloid and Interface Science* **86**, 196–203.

KRISTOFFERSEN, A., ROLLA, G., SONJU, T. & JANTZEN, E. 1982b The organic film developed on metal surfaces exposed to seawater: chemical studies. *Journal of Colloid and Interface Science* **86**, 191–196.

LOEB, G. J. & NEIHOF, R. A. 1975 Marine conditioning films. *Advances in Chemistry Series* **145**, 319–335.

MARSHALL, K. C. 1973 Mechanism of adhesion of marine bacteria to surfaces. *Proceedings of the 3rd International Congress on Marine Corrosion and Fouling*, eds Acker, R. F. *et al.*, pp. 625–632. Evanston, Ill: Northwestern University Press.

OBUEKWE, C. O., WESTLAKE, D. W. S., COOK, F. D. & COSTERTON, J. W. 1981 Surface changes in mild steel coupons from the action of corrosion-causing bacteria. *Applied and Environmental Microbiology* **41**, 766–774.

PEDROS-ALIO, C. & BROCK, T. D. 1983 The importance of attachment to particles for planktonic bacteria. *Archives of Hydrobiology* **98**, 354–379.

QUINN, J. P. 1984 The modification and evaluation of some cytochemical techniques for the enumeration of metabolically active heterotrophic bacteria in the aquatic environment. *Journal of Applied Bacteriology* **57**, 51–57.

ROSSER, H. R. & HAMILTON, W. A. 1983 Simple assay for accurate determination of [^{35}S]sulfate reduction activity. *Applied and Environmental Microbiology* **45**, 1956–1959.

RUSESKA, I., ROBBINS, J. & COSTERTON, J. W. 1982 Biocide testing against corrosion-causing oil-field bacteria helps control plugging. *Oil and Gas Journal*, March, 253–264.

SAKAGUCHI, I. 1980 Studies on the biofouling and corrosion of condensers in steam power plants in Japan. *Proceedings of the Symposium on Condenser Biofouling and Control*, eds Garey, J. F. *et al.*, pp. 77–86. Michigan: Ann Arbor Science.

SANDERS, P. F. & MAXWELL, S. 1983 Microfouling, macrofouling and corrosion of metal test specimens in seawater. *Microbial Corrosion. Proceedings of the Conference on Microbial Corrosion, 8–10 March 1983, NPL, Teddington*, pp. 74–83. London: The Metals Society.

SMITH, A. D. 1982 Immunofluorescence of sulphate-reducing bacteria. *Arvhives of Microbiology* **133**, 118–121.

TABOR, P. S. & NEIHOF, R. A. 1982 Improved method for determination of respiring individual microorganisms in natural waters. *Applied and Environmental Microbiology* **43**, 1249–1255.

TANNER, R. S., HAACK, T. K., SEMET, R. F. & GREENLEY, D. E. 1985 A mild steel tubular flow system for biofilm monitoring. In press.

TARBUCK, L. A. & WYBORN, C. H. E. 1980 Microbial growth and activity in condenser tubes. *Proceedings of the Symposium on Condenser Biofouling and Control*, eds Garey, J. F. *et al.*, pp. 85–103. Michigan: Ann Arbor Science.

TERRY, L. A. & EDYVEAN, R. G. J. 1984 Influences of microalgae on corrosion of structural steel. *Corrosion and Marine Growth on Offshore Structures*, eds Lewis, J. R. & Mercer, A. D., pp. 38–44. Chichester: Ellis Horwood.

TILLER, A. K. 1983 Electrochemical aspects of microbial corrosion: An overview. *Microbial Corrosion. Proceedings of the Conference on Microbial Corrosion, 8–10 March 1983, NPL, Teddington*, pp. 54–65. London: The Metals Society.

TRULEAR, M. G. & CHARACKLIS, W. C. 1982 Dynamics of biofilm process. *Journal of the Water Pollution Control Federation* **54**, 1288–1301.

WATSON, S. W., NOVITSKY, J. J., QUINBY, H. L. & VALOS, F. A. W. 1977 Determination of bacterial number and biomass in the marine environment. *Applied and Environmental Microbiology* **33**, 940–946.

ZAIDA, B. R., BARD, R. F. & TOSTESON, T. R. 1984 Microbial specificity of metallic surfaces exposed to ambient seawater. *Applied and Environmental Microbiology* **48**, 519–524.

ZAMBON, J. J., HUBER, P. S., MEYER, A. E., SLOTS, J., FORNALIK, M. S. & BAIER, R. E. 1984 *In situ* identification of bacterial species in marine microfouling films using immunofluorescence techniques. *Applied and Environmental Microbiology* **48**, 1214–1220.

ZOBELL, C. E. 1943 The effect of solid surfaces on bacterial activity. *Journal of Bacteriology* **46**, 39–56.

ZOBELL, C. E. 1972 Substratum as an environmental factor for aquatic bacteria, fungi and blue green algae. *Marine Ecology*, Vol. 1. *Environmental Factors*, ed. Kinne, O., pp. 1249–1272. New York: Wiley & Sons Inc.

ZSOLNAY, A. & LITTLE, B. 1983 Characterization of fouling films by pyrolysis and chemical ionization mass spectrometry. *Journal of Analytical and Applied Pyrolysis* **4**, 335–341.

The Laboratory Generation of Biofilms and their Use in Biocide Evaluation

P. N. GREEN, I. J. BOUSFIELD AND ANGELA STONES
National Collections of Industrial and Marine Bacteria Ltd, Torry Research Station, P.O. Box 31, 135 Abbey Road, Aberdeen AB9 8DG

The way in which microbial populations develop on submerged surfaces is becoming the subject of increasing attention (Wardell *et al.* 1983). Under suitable conditions, clean surfaces rapidly become coated with a complex, often strongly adherent biofilm or glycocalyx (Costerton *et al.* 1978) consisting of microorganisms embedded in a matrix of organic material. In many industrial process systems the unwanted development of such biofouling films and the activities of the organisms in them are a serious nuisance. A well-known example is the fouling of water injection systems in the North Sea oil industry by sulphate-reducing bacteria (SRB) and other bacteria, leading to corrosion of pipework and other problems. Although in the natural marine environment, the activities of SRB are confined largely to sediments, these bacteria occur in significant numbers throughout the water column (Crouch 1983) and thus are present in the sea water used in injection systems. The highly anaerobic conditions in these systems and the production of metabolizable substrates by facultative organisms create an ideal environment for the growth of SRB which, if not controlled, can cause severe problems, e.g. corrosion, reservoir plugging, oil souring and toxic H_2S production (see e.g. Hamilton & Sanders 1984).

Injection systems are commonly treated with biocidal chemicals to inhibit the development of SRB-containing biofilms, but such treatments are often far from satisfactory (Wilkinson 1982). Several laboratory studies have shown that organisms embedded in a complex biofilm are often protected from the lethal effects of various chemical agents, largely because these agents penetrate the outer layers of the matrix only to a limited extent (Drysdale 1968; Costerton *et al.* 1981; Gaylarde & Johnston 1983). Thus the traditional method of assessing a biocide by measuring its activity against a suspension of organisms in pure culture can have little relevance in such situations.

Industrial
Microbiological Testing

0–632–01793–7

Laboratory Biofilm Generators

Robbins Device

In an attempt to develop a more meaningful method for the evaluation of biocides against sessile populations, McCoy *et al.* (1981) and McCoy & Costerton (1982) devised a laboratory rig for the generation of biofilms. This 'Robbins device' consisted of a length of brass tubing containing removable studs inserted at intervals along its length so that their inner surface was flush with the inner wall of the pipe. The studded pipe formed part of a closed loop through which a mixed bacterial culture enriched from seawater was continually circulated until a biofilm had developed on the surface of the studs. Biocide was then circulated through the device and individual studs were removed at intervals so that the effect of the biocide on their adherent biofilm could be assessed.

NCIMB Biofilm Generator

The original 'Robbins device' was made from Admiralty brass. To reproduce it would require expensive metal engineering skills and workshop facilities. For our biocide testing programme we needed test rigs that were both cheap and easy to build. Therefore we developed the NCIMB biofilm generator, which, with the exception of the removable studs, is built from standard plumbing and laboratory materials.

The biofilm generator consists of a 2 m length of high-density polyethylene pipe (Hayes polythene, class C, obtainable from any plumber) of internal diameter 18.75 mm (0.75 in.). A flat surface *c.* 10 mm wide is made along the length of the pipe with a hand plane. Holes (60 × 6 mm) are drilled through the wall of the pipe at *c.* 25 mm intervals along the centre line of this flat surface, leaving *c.* 250 mm undrilled at each end of the pipe. Mild steel studs (made to order by Edgar Engineering Services Ltd, Ayr, Scotland, to the dimensions shown in Fig. 2) fitted with nitrile O-rings (6 mm i.d., 1.75 mm section) are inserted into the holes and held tightly in place with no. H30 (30 mm) Unex hose clips (Uniclip Ltd, Byfleet, Weybridge, U.K.) so that the O-rings are firmly compressed against the flat surface of the pipe (Fig. 1). The studded pipe is supported on a framework of laboratory scaffolding with the row of studs at an angle of *c.* 45° from the vertical and is connected via 18.75 mm (0.75 in.) PVC connectors (FIP fittings, obtainable from plumbers' merchants) and PVC or polyethylene tubing to a polypropylene, magnetically coupled, centrifugal, circulating pump (CP Instruments, Bishops Stortford, U.K.; Cat. No. CR 1034-OE) to form a recirculating loop of total volume *c.* 2 litres. Inlet and outlet ports consist of PVC T-connectors fitted with rubber bungs carrying glass/PTFE high performance vacuum stopcocks (Quickfit Rotaflo

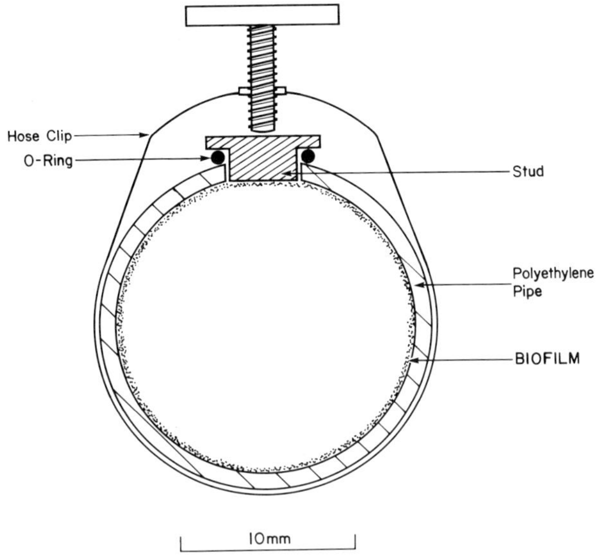

FIG. 1. Cross section (through a stud) of the pipe of the NCIMB biofilm generator.

HP2/6). Flow rate is controlled by a 15 mm (0.5 in.) PVC ball valve (FIP fitting from plumbers' merchants) and is set initially by temporarily incorporating a water flowmeter (Neptune Measurement Ltd, Oldham, U.K.; Type T10) into the circuit and circulating clean water. The flowmeter is not kept permanently in the circuit because it is susceptible to corrosion by seawater. The layout of the various components is shown in Fig. 3 and photographs of the rig and stud insertion are shown in Figs. 4 and 5.

Media Employed in Test Procedure

Modified API medium (American Petroleum Institute API RP38, 1975)
The composition is as follows: sodium lactate (70% soln) 4 ml; yeast extract, 0.1 g;

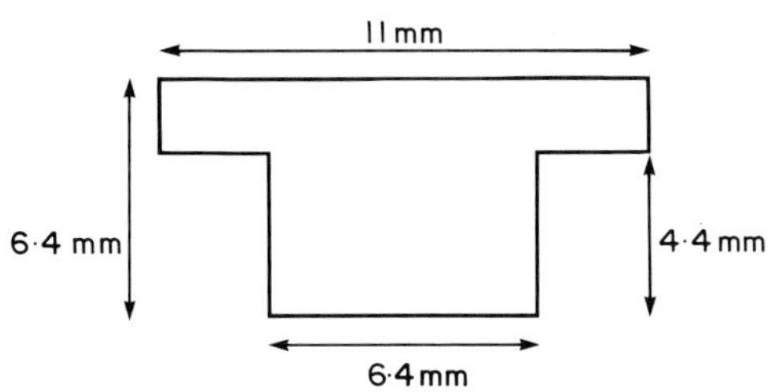

FIG. 2. Studs used in the NCIMB biofilm generator.

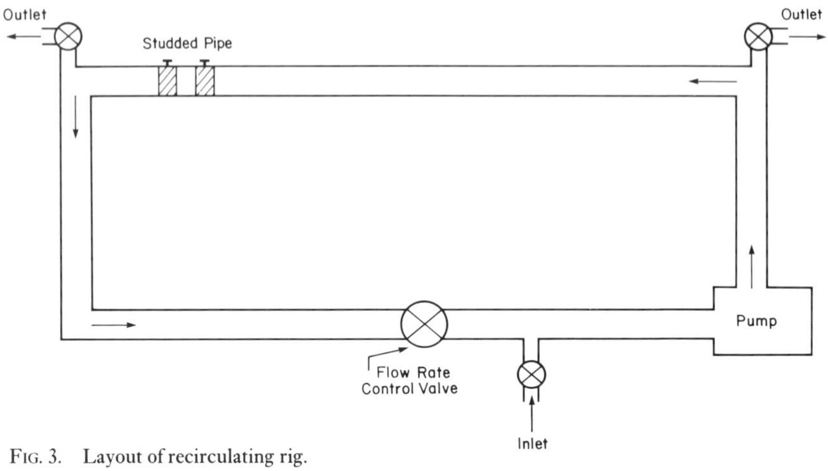

FIG. 3. Layout of recirculating rig.

FIG. 4. The NCIMB biofilm generator (rig).

ascorbic acid, 1.0 g; KH_2PO_4, 0.1 g; $(Fe(SO_4)_2(NH_4)_2.6H_2O$, 0.1 g; Resazurin (0.1% (w/v) aqueous solution), 1 ml; filtered seawater, 1 litre, pH to 7.2–7.5. Quarter strength and one-tenth strength medium have the same composition but contain 1 ml and 0.4 ml sodium lactate solution respectively.

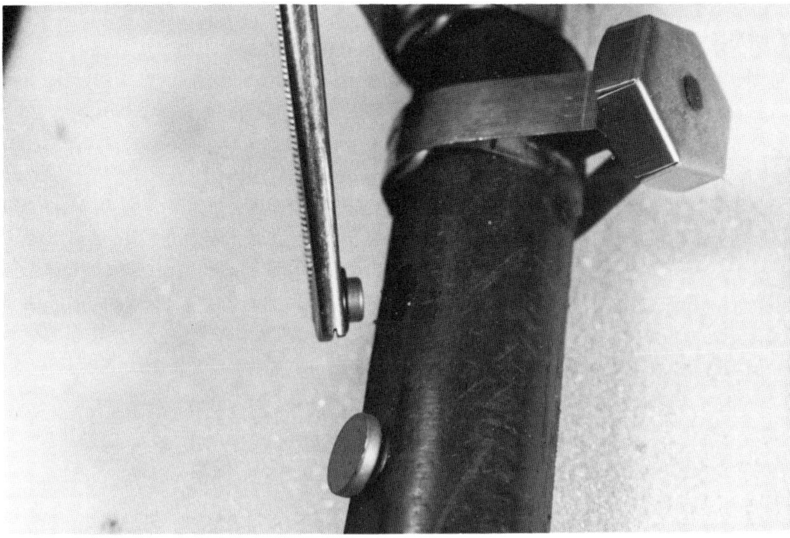

FIG. 5. Insertion of stud into biofilm generator.

Seawater yeast peptone medium

The composition was as follows: yeast extract, 3 g; peptone, 5 g; aged, filtered seawater (seawater is stored in the dark for 3 weeks (minimum) to 'age' it), 750 ml; tap or distilled water, 250 ml; pH 7.3.

Biofilm Generation

A mixed culture of facultative and sulphate-reducing bacteria enriched from a North Sea water injection system is grown up in 250 ml/full strength modified API medium for 5 days at 30°C. This culture is then added to 10 litre quarter-strength modified API medium treated, prior to inoculation, as follows: whilst the broth is still hot from autoclaving, sparging with oxygen-free nitrogen (OFN) is begun and is continued until the broth is no longer pink and has cooled to room temperature. After incubation for 3 days at 30°C the culture is ready for use. Before culture is introduced into the biofilm generator, the entire recirculating system is pressurized with OFN then isolated from the OFN supply and left for *c.* 0.75 h. The sound of escaping gas on opening an outlet valve means that a positive pressure has been maintained and the system is therefore assumed to be gastight. Failure to maintain a positive pressure indicates that the system leaks, whereupon all studs and joints are checked for tightness and the test is repeated. When leak proof, the system is filled by blowing in culture with OFN. Inlet and outlet valves are then closed and the culture is circulated continuously through the biofilm generator at a velocity of *c.* 0.8 m/s. At intervals during the first day of operation, the pump is turned off and residual OFN is bled from the system, which becomes completely degassed within this 24 h. Thereafter *c.* 1 litre amounts of liquid are displaced daily from the system with culture from the 10 litres reservoir. The reservoir is replenished weekly with fresh culture. Effluent culture from the rig is collected in a vessel vented to the atmosphere via a copper sulphate (5% aq. solution) trap to remove H_2S. After the first week of operation, cultures in one-tenth strength modified API broth are used to feed the system. At regular intervals, studs are pulled randomly from the biofilm generator and their surface populations counted (see below) until satisfactory and consistent counts are obtained (up to 4 weeks) and an even, black biofilm coats the studs (Figs. 6 and 7). A typical biofilm would contain *c.* 10^7 SRB and 10^6 other bacteria per cm^2. Each stud removed from the rig is replaced immediately with a fresh stud, and during stud removal and replacement, the pump is turned off.

Procedures for Counting Bacteria on Stud Surface

Each stud is transferred to 10 ml sterile seawater and placed immediately in an anaerobic cabinet containing an atmosphere of N_2:H_2:CO_2 (80:10:10 (by volume)). The biofilm is removed by scrubbing the stud with a small pre-sterilized piece (*c.* 1 cm^2) of plastic wool teased from a domestic scouring pad and the resulting suspension is shaken vigorously by hand before being diluted serially in decimal steps to 10^{-8}. This dilution series is then removed from the cabinet and used to inoculate 9 ml amounts of modified API medium in triplicate by means of

FIG. 6. Stud coated with biofilm (L) and clean stud (R).

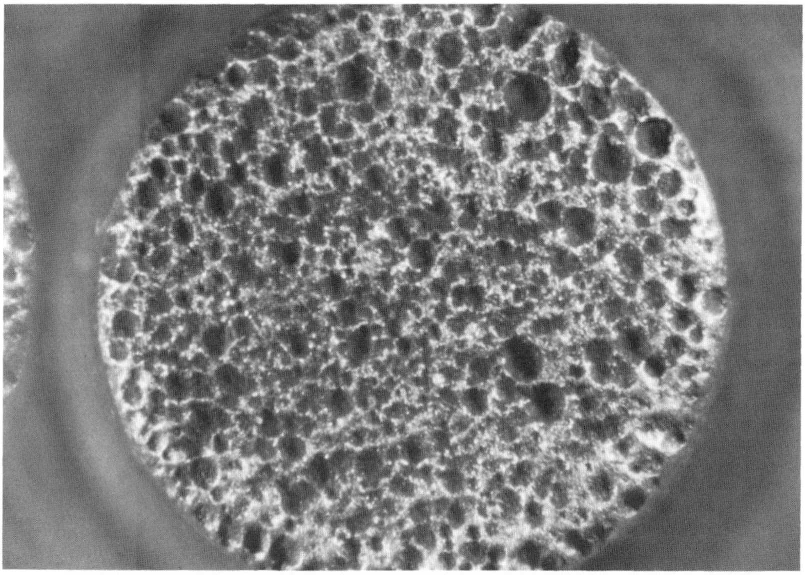

FIG. 7. Surface of stud showing pitting caused by the biofilm (*c.* × 25).

hypodermic syringes thrust through the septa of the bottle caps to maintain anaerobiosis. The same dilution series is then used to prepare aerobic plate counts. Amounts (100 μl) of each dilution are spotted in triplicate on to the surface of pre-dried seawater–yeast–peptone plates.

The bottles of modified API medium are incubated at 30°C and examined for up to 14 days for blackening, indicating the growth of SRB, after which the most probable numbers (MPN) and confidence limits of SRB are estimated using a computer program written by G. L. Smith of the Torry Research Station. Aerobic plates are incubated at 25°C and colonies are counted after 3 and 5 days.

Biocide Evaluation

There are two ways of testing biocides against biofilms developed on the surface of studs, which we will refer to as dynamic and static testing.

Dynamic Testing

Control studs are first pulled from the rig to allow determination of the numbers of organisms present in the biofilm before the addition of biocide. Biocide is then introduced into the recirculating system to give a particular concentration and is allowed to circulate for a predetermined period (contact time), e.g. 6 h, which is a contact time frequently used when biocides are being applied in offshore situations. At intervals during the contact time, studs are pulled and the numbers of viable organisms remaining in the biofilm are estimated as described above. The advantage of dynamic testing is that the performance of the biocide is assessed in a flowing system which to some extent is analogous to the field situation. The major drawback is that only one dosage regime of a single or mixed biocide can be tested at a time. Comparative dynamic testing requires either several biofilm generators operating in parallel or intervals of 1–4 weeks between tests to allow the generation of new biofilms. Nevertheless, in principle, dynamic testing is the preferred approach.

Static Testing

Studs are removed from the rig for exposure of the biofilms to biocide. A freshly pulled stud is suspended by a wire carrier in 100 ml sterile seawater containing biocide at the appropriate concentration (Fig. 8). Throughout the contact time the solution is stirred magnetically and is sparged with OFN to exclude air. A separate control stud is exposed to sterile seawater in parallel with the test stud for the same contact time.

A reference biocide (25% glutaraldehyde at 0.02% v/v for 6 h) is also included in each test run so that in the event of variation between biofilms generated at different times, the performance of test biocides can be assessed in relation to a common standard. After exposure to biocide for a selected contact time all studs are transferred immediately to 10 ml amounts of sterile seawater and bacterial counts are done as described above.

Although static testing resembles field conditions less than does dynamic testing, it has the great advantage that several biocides can be tested simultaneously at a number of concentrations and contact times. It is thus of great value in screening biocides to select those that are worthy of dynamic testing and in determining optimum combinations of concentration and contact times. Knowledge of the latter is particularly important since some biocides are effective at high concentrations for short contact times while others perform better at lower concentrations over longer periods. For example, glutaraldehyde may exert its full lethal effect within half an hour, whereas the isothiazolones may take several hours to achieve a similar effect, but at much lower concentrations. It is also important to realize that there are limits to the maximum effective concentrations of most biocides, beyond which further increases in concentration are merely wasteful. Excessive dosing with biocides is, unfortunately, not uncommon in the field. Similarly, contact times in practice tend often to be determined with more regard to convenient working practices than to the properties of the particular biocide being used. However, experimenting with different dosing regimes in the field situation, particularly offshore, can be inconvenient, difficult and expensive. Therefore, the development of laboratory

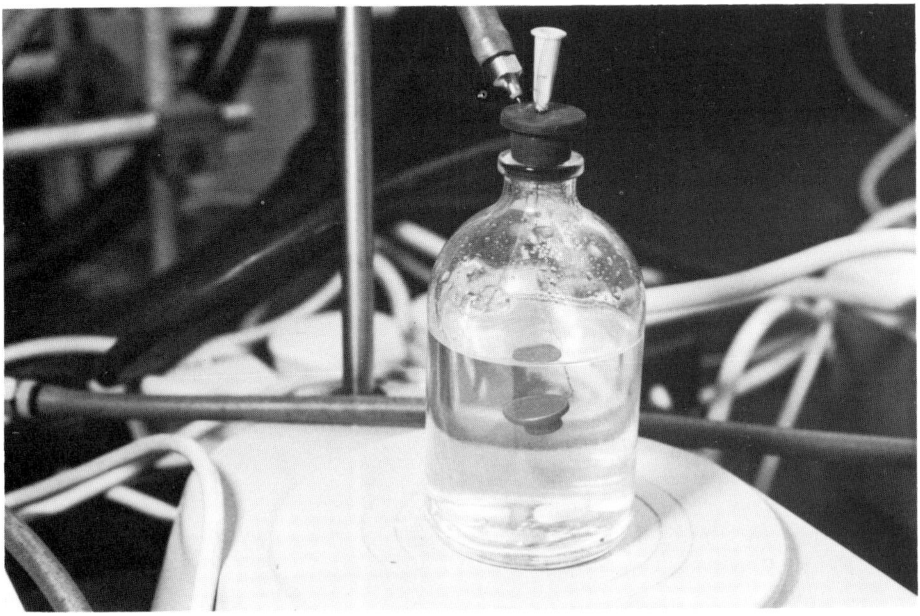

FIG. 8. Static testing: exposure of fouled stud to biocide.

methods for evaluating biocides under conditions resembling those found in the field is of great importance. Although the methods described in this chapter are at best a crude representation of the real situation, they are nevertheless a step towards obtaining more meaningful data than can be obtained using traditional laboratory methods of assessing antimicrobial chemicals against free-floating populations.

References

AMERICAN PETROLEUM INSTITUTE 1975 *Recommended Practice for Biological Analysis of Subsurface Injection Waters*, 3rd edn. API RP 38. Dallas: Division of Production, American Petroleum Institute.

COSTERTON, J. W., GEESEY, G. G. & CHENG, K. J. 1978 How bacteria stick. *Scientific American* **238**, 86–95.

COSTERTON, J. W., IRVIN, R. T. & CHENG, K. J. 1981 The bacterial glycocalyx in nature and disease. *Annual Review of Microbiology* **35**, 299–324.

CROUCH, B. 1983 Evaluation of biocides for use in North Sea Oil operations. *Biodeterioration 5*, eds Oxley T. A. & Barry, S., pp. 725–731. Chichester: John Wiley & Sons.

DRYSDALE, G. R. 1968 Chlorine and chlorine compounds. *Disinfection, Sterilisation and Preservation*, eds Lawrence, C. A. & Block, S. S., pp. 278–304. Philadelphia: Lea & Febiger.

GAYLARDE, C. C. & JOHNSTON, J. M. 1983 The effect of some environmental factors on biocide sensitivity in *Desulfovibrio*: implications for biocide testing. *Microbial Corrosion*, pp. 91–97. London: The Metals Society.

HAMILTON, W. A. & SANDERS, P. F. 1984. Microbial corrosion: experience in the offshore oil industry. *Biodeterioration* 6 (in press).

MCCOY, W. F., BRYERS, J. D., ROBBINS, J. & COSTERTON, J. W. 1981 Observations of fouling biofilm formation. *Canadian Journal of Microbiology* 27, 910–917.

MCCOY, W. F. & COSTERTON, J. W. 1982 Fouling biofilm development in tubular flow systems. *Developments in Industrial Microbiology* 23, 551–558.

WARDELL, J. N., BROWN, C. M. & FLANNIGAN, B. 1983 Microbes and surfaces. *Microbes in their Natural Environments* (S.G.M. Symposium 34), eds Slater, J. H., Whittenbury, R. & Wimpenny, J. W. T., pp. 351–378. Cambridge University Press.

WILKINSON, T. J. 1982 Control of bacterial growth: the use of biocides and alternative control systems in the offshore oil industry. *Biocides in the Oil Industry*, pp. 39–68. Institute of Petroleum Symposium. Chameleon Press Ltd.

Testing for Selective Toxicity and Mutagenicity against the Mitochondrion in *Saccharomyces cerevisiae*

IAN J. BRUCE, IVOR H. EVANS AND DAVID WILKIE*
Department of Biological Sciences and Environmental Health, Thames Polytechnic, Wellington Street, London SE18 6PF, and
**Department of Botany and Microbiology, University College London, Gower Street, London WC1E 6BT*

Introduction

(i) *Yeast as a model eukaryotic cell*
Microbial prokaryotes are now well established test organisms for the assessment of the potential toxicity and carcinogenicity of chemicals, and experimental protocols usually require activation systems such as mammalian liver microsome preparations. A substance's 'genotoxicity' is inferred from its effectiveness in inducing point mutations in the bacterial chromosome (Venitt & Parry 1985).

However, the huge differences in cell organization between pro- and eukaryotes obviously limit the usefulness of bacteria in assessing probable sites and mechanisms of action of toxicants in the eukaryotic cell. By contrast, yeast (*Saccharomyces cerevisiae*), the best understood of all eukaryotic microbes (Strathern *et al.* 1981, 1982) is an organism which contains its own activation systems (Woods & Wiseman 1980) and which offers a variety of means for detecting genotoxicity, all of which should be relevant to higher organisms. Yeast can also be used as a model eukaryotic cell (Herskowitz 1985) to identify subcellular targets and to dissect mechanisms of action (Wilkie 1972). Yeast's amenability to classical genetic procedures, as well as the relative ease with which yeast genes can be isolated and manipulated by recombinant DNA techniques (Struhl 1983) should generate highly sophisticated new approaches to the study of cell–toxicant interactions: it will be possible to construct strains with very precisely determined lesions or alterations in gene products concerned in the intoxication process, so enhancing the sensitivities of mutagenicity test procedures and offering more rigorous examination of postulated mechanisms.

Industrial
Microbiological Testing

0-632-01793-7

(ii) *The mitochondrion as a target*

Many xenobiotics have selective activity against organelles and systems within the eukaryotic cell. One such target organelle is the mitochondrion, and *S. cerevisiae* is particularly well suited to detecting and analysing primary anti-mitochondrial activity of compounds as varied as: tricyclic anti-depressants (Linstead & Wilkie 1971), non-steroidal anti-inflammatory drugs (Bruce & Wilkie 1983, 1984), carcinogens (Egilsson *et al.* 1979) and antineoplastics (Evans *et al.* 1973). Methods and approaches available in this area of research, using yeast as a model test organism, are summarized in this article.

S. cerevisiae is a facultative anaerobe and so is unusual among eukaryotes in being able to grow and proliferate in the absence of a functional respiratory system, provided an adequate supply of a suitable fermentable carbon source, such as glucose, is available. It is this property which enables *primary* anti-mitochondrial activity to be detected so readily in yeast. Agents inhibiting mitochondrial biogenesis or key mitochondrial functions such as electron transport or oxidative phosphorylation would be highly destructive and ultimately lethal to most eukaryotic cells, precipitating many secondary effects obscuring the primary anti-mitochondrial action. However, in yeast, an inhibitor primarily active against the mitochondrial system has little effect on fermentative growth. An additional attraction of *S. cerevisiae* in this area of work is that the yeast mitochondrion is very well understood at a molecular genetic level, as yeast has been the foremost experimental organism in the field of mitochondrial genetics.

Organelles as large and complex as the mitochondrion clearly contain many unique features which could constitute sites of action of selectively anti-mitochondrial agents. Major mitochondrial targets and possible routes of toxicant action are illustrated in Fig. 1.

Methodology

(1) *Growth Tests to Detect Primary Anti-mitochondrial Activity*

A first indication of anti-mitochondrial activity can be obtained by comparing an agent's ability to inhibit yeast growth on media containing either a fermentable (e.g. dextrose) or non-fermentable (e.g. glycerol, succinate) carbon source. The first tests are conducted on solid medium using a genetically varied array of yeast strains, the fermentable medium being YED, containing 1% yeast extract, 2% dextrose and 2% agar and the non-fermentable medium being YEG, containing 1% yeast extract, 4% (w/v) glycerol and 2% agar. A supplement of 2% peptone can be added to both media. The yeast strains are inoculated on to the two media containing identical ranges of toxicant concentrations using a variety of drop inoculation techniques.

In one system, twenty-seven oxoid caps are cemented into a glass petri dish using epoxy resin or, for speed and simplicity, 'Blu-Tack' (J. F. R. Spencer, pers. commun.); these provide wells for suspension in sterile distilled water of the various strains. Alternatively microtitre plates can be used to provide a larger number of sterile wells. In both cases ethanol-flameable mild steel rod inoculators are constructed with prongs congruent with the wells and able to fit into a standard petri dish (Fig. 2).

After 2 days incubation at 30°C, growth is scored, semiquantitatively by eye. An example of such a test, for aspirin, is given in Table 1.

Differences in minimal inhibitory concentration (MIC) on YEG and YED of two-fold or more, are considered significant, a greater sensitivity to growth

TABLE 1. *The minimum inhibitory concentrations of aspirin in YED and YEG medium for eighteen* S. cerevisiae *strains*

	Minimum inhibitory concentration (mg/ml)	
Strain	YED	YEG
D11	4.0	3.0
A30	4.0	3.0
188 wt	4.0	2.5
B21	4.0	2.5
6/81	4.0	2.5
Ber B	4.0	3.0
22/701	4.0	3.0
D18	4.0	2.5
41/161	4.0	3.0
B27	4.0	2.5
D22	4.0	2.5
2180A	4.0	2.5
2180a	4.0	2.5
D4	4.0	2.5
B41	4.0	2.5
A285	4.0	2.5
Ber A	4.0	2.5
D6	4.0	2.5
Average	4.0	2.64

Eighteen *S. cerevisiae* strains were grown to stationary phase on YED medium and cells from each strain drop inoculated simultaneously onto a doubling concentration series of aspirin in YED and YEG medium (see methodology). The plates were incubated at 30°C for 3 days and subsequently scored visually for growth.

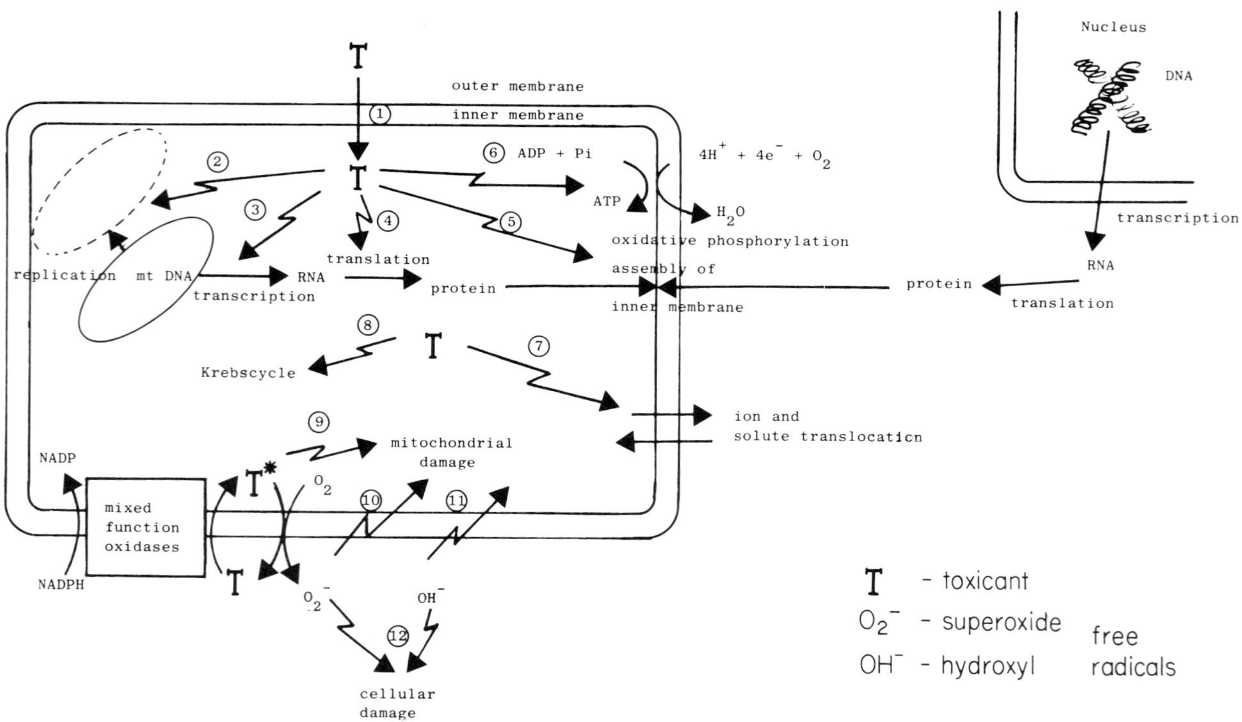

FIG. 1. The mitochondrion as a toxicant target.

(1) Mitochondria actively accumulate many ions and metabolites as part of their normal biosynthetic and bioenergetic functions. Toxicants (T) with structural similarities to these compounds may well be accumulated and sequestered within the mitochondrion. Dimethyl-nitrosamine and the phenanthridine dye ethidiam bromide are both actively accumulated by mitochondria.

The inner mitochondrial membrane also provides an extensive hydrophobic phase which in some cell types must comprise a substantial fraction of the entire cell's lipoprotein complement. Hydrophobic toxicants such as polycyclic hydrocarbons may accumulate in the inner membrane by simple partitioning.

(2) Mitochondrial DNA (mtDNA) is a small, usually covalently closed circular DNA species with a limited coding capacity but a high copy number. *S. cerevisiae* and mammalian mtDNAs (approximately 75 Kbp and 15 Kbp respectively), though differing in size, have a similar content of genetic information mainly encoding 2 rRNA species, 20–30 tRNA species and about 10 inner membrane proteins (e.g. the apoprotein of cytochrome b and polypeptides of the cytochrome oxidase and ATPase enzyme complexes). mtDNA, not being packaged into a nucleoprotein chromatin structure, may be especially vulnerable to DNA reactive compounds, such as intercalators, alkylators, and others. In mammalian mtDNA especially, there is virtually no non-coding material, so that mutational damage is more likely to result in an altered or impaired gene product than in the case of nuclear DNA. mtDNA replication occurs throughout the cell cycle and repair of mtDNA is probably less efficient than that of nuclear DNA. Mitochondrial mutations in *S. cerevisiae* can result in characteristic mutant phenotypes such as respiratory deficiency (*petite*, mit^- see below) and antibiotic resistance.

(3) DNA-binding substances inhibit mtDNA replication and also tend to inhibit transcription, but no specific inhibitors of mtRNA polymerase are yet known.

(4) Mitoribosomes are quite distinct from the cytoplasmic species and show antibiotic sensitivity to chloramphenicol and erythromycin similar to prokaryotic ribosomes. Both these compounds are inhibitors of mitochondrial translation. The mitochondrial translation mechanism differs in many respects (e.g. initiation, tRNAs, amino-acyl tRNA synthetases) from that in the cytoplasm and this may be reflected in differential susceptibilities to toxicants such as amino acid analogues and folate antagonists.

(5) Growth of the inner mitochondrial membrane requires coordinated incorporation of proteins synthesized both intra- and extra-mitochondrially. Most ($>90\%$) mitochondrial proteins are specified by nuclear genes and synthesized in the cytoplasm. Many of these proteins have leader sequences and translocation into and across the inner mitochondrial membrane requires precise proteolytic clearages.

(6) A major mitochondrial function is phosphorylation of ADP to ATP using respiratory energy. This process of oxidative phosphorylation is acutely dependent upon the integrity of the inner membrane, where many components of the electron transport chain are located. Many specific inhibitors acting at discrete sites in this multi-component ensemble are known, e.g. (a) antimycin, which blocks the passage of electrons from cytochrome b to cytochrome c_1; (b) cyanide which blocks the passage of electrons from cytochrome oxidase to the terminal acceptor oxygen; and (c) oligomycin which inhibits phosphorylation, binding to a subunit of the mitochondrial ATPase. Numerous 'uncouplers' are known, all probably acting by discharging the trans-membrane proton gradient.

(7) Translocation of ions and solutes across the inner mitochondrial membrane is susceptible to a number of specific inhibitors. For example, valinomycin, an ionophore, potentiates potassium ion transport, while calcium ion transport is inhibited by ruthenium red and stimulated by certain local anaesthetics. Adenine nucleotide translocation is blocked by bongkrekic acid.

(8) The mitochondrial matrix harbours numerous soluble enzymes which may be specifically inhibited by toxicants.

(9), (10) and (11) *S. cerevisiae* contains a cytochrome p448/p450 mixed function oxidase system very similar to that of mammals (probably located in the endoplasmic reticulum). Mixed function oxidase systems able to metabolize and activate drugs also occur in mitochondrial preparations. These enzyme systems, aimed at detoxification of xenobiotics, use molecular oxygen and NADPH to oxidize (usually) hydrophobic substrates to more electrophilic species. Reactive species can be generated able to interact directly with and damage neighbouring macromolecules, or pass electrons to oxygen, so producing highly damaging forms such as superoxide, hydroxyl free radicals, and singlet oxygen.

(12) The damage inflicted by these free radicals could be both intra- and extra-mitochondrial.

(13) Any agents able selectively to alter the relevant nuclear genes or their role in the production of mitochondrial component, could have a selective anti-mitochondrial effect, though it is difficult to conceive how purely chemical toxicants could achieve such specificity.

The general topic of modes of mitochondrial intoxication is discussed in Egilsson *et al.* (1979) and Wilkie & Evans (1982) and more extensively in Wilkie *et al.* (1983); numerous references are given. The molecular genetics of yeast mitochondria are reviewed at length in Evans (1983).

FIG. 2. A multiple inoculator system for yeast strains. The inoculator consists of twenty-seven brass rods mounted in a perspex block. The rods are congruent with wells (approximately 1.5 ml capacity) drilled in an aluminium block, inoculation from which yields an array of twenty-seven patch cultures on a standard petri dish.

inhibition on YEG indicating a primary anti-mitochondrial effect. Strains showing the greatest contrast in sensitivities would be chosen for further study.

Confirmation and quantification of the initial results is achieved by repeating the growth inhibition tests on selected strains in liquid medium, monitoring growth by optical density (O.D.) or dry weight measurements (Erlenmeyer side arm flasks which are able to fit directly into a colorimeter are convenient for O.D. measurements). Figs. 3(a) and 3(b) give an example of such a liquid medium growth test experiment.

(2) *Oxygen Electrode and Related Studies*

Differences in MIC on fermentable and non-fermentable media due to trivial causes, such as differences in cell permeability to the inhibitor on the two media, are excluded by studies of cells grown on fermentable medium in the presence of inhibitor concentrations which would suppress growth on non-fermentable medium. Reduced respiration in inhibitor-treated cells indicates that mitochondria are indeed the target. Direct inhibition of the respiration of whole cells or of isolated mitochondria implies an action on a mitochondrial function such as electron transport or phosphorylation of ADP.

(i) *Whole cell studies*

Cells are grown with vigorous aeration in liquid fermentable medium, in the presence and absence of the inhibitor, into stationary phase, the respiration of the control cells being thoroughly derepressed. The respiratory activity of the cells can be readily assessed using any of a variety of commercially available oxygen electrodes (e.g. Clark type, Rank Ltd) connected to a flat bed chart recorder (Bruce 1983; Estabrook 1967). Direct effects of the growth inhibitor on control cells actively

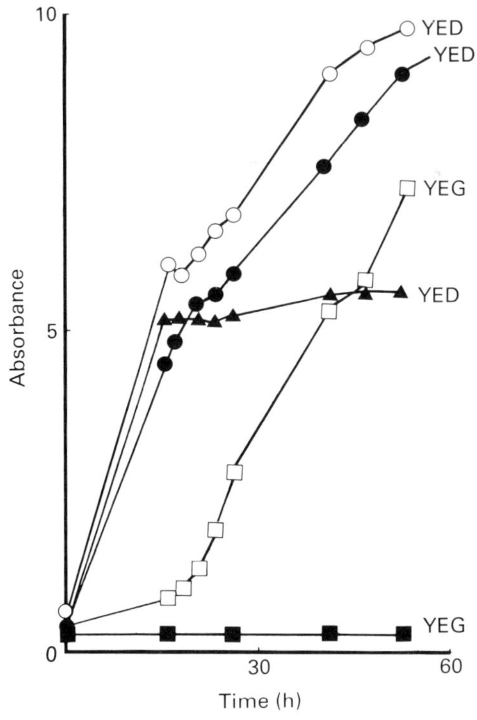

FIG. 3(a). Growth of strain A285 in glucose (YED) or glycerol (YEG) liquid media, monitored using side-arm shake flasks and an EEL colorimeter with appropriate blanks. Flasks contained no drugs (○), adriamycin at 2 μg/ml (2), or erythromycin at 1 mg/ml (ER).

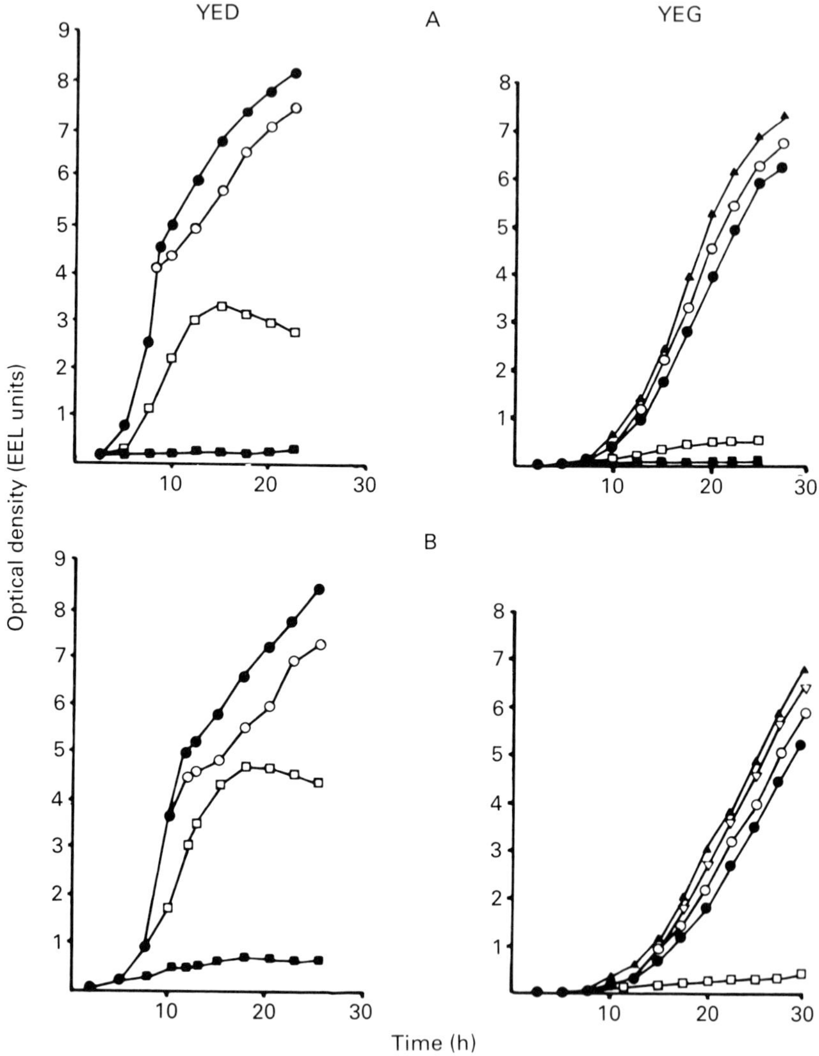

FIG. 3(b). The effect of aspirin on the growth of strains Ber B and A30 in liquid YED and YEG media. Growth strains Ber B and A30 in glucose (YED) or glycerol (YEG) media monitored using side-arm shake flasks and an EEL colorimeter with appropriate blanks. Flasks contained aspirin at 1 mg/ml: (A) YED: ●, control, 0.05, 0.10, 0.50; ○, 1.00; □, 2.00; ■, 3.00. YEG: ●, control; ▲, 0.05, 0.10, 0.50; ○, 1.00; □, 2.00; ■, 3.00. (B) YED: ●, control, 0.05, 0.10, 0.50; ○, 1.00; □, 2.00; ■, 3.00. YEG: ●, control; ○, 0.05, 1.00; ▽, 0.10; ▲, 0.50; □, 2.00, 3.00.

respiring in the electrode chamber can also be observed. The effects of various concentrations of adriamycin and aspirin on oxygen uptake of yeast cells can be seen in Figs. 4(a) and 4(b).

(ii) *Studies on isolated mitochondria*
Functional inhibition of mitochondria may not be demonstrable in whole cells, for example, because of slow permeation. The problem is avoided by observing the agents' effects on isolated mitochondria *in vitro.*

The thick cell walls of *S. cerevisiae* present problems in the isolation of intact subcellular organelles, but it is possible to make yeast mitochondrial preparations with oxidation and phosphorylation sufficiently 'coupled' for useful *in vitro* studies.

A simple and rapid isolation protocol employing mechanical disruption is based on that of Spencer *et al.* (1971). Cells are grown to late log phase in a medium containing 1% yeast extract (Difco), 2% peptone (Difco 'Bacto'), 1% glucose, and 1% ethanol, with vigorous aeration at 30°C. Cells are harvested by centrifugation and washed twice with cold distilled water. The final cell pellet is weighed (10 g wet weight produces an adequate yield of mitochondria) and resuspended at a concentration of 1 g/ml in isolation medium (0.25 M sucrose, 0.02 M KH_2PO_4, 0.001 M EDTA (ethyleneglycol-bis-(β-aminoethylether)N,N-tetraacetic acid), 0.01 M $MgCL_2$, pH 6.8). This suspension is mixed with Ballotini beads (Jencons, 0.45–0.50 mm diameter), 5 g beads/g cells, and disrupted by shaking at full speed for 15 s in a CO_2-cooled Braun (B. Braun, Melsungen) homogenizer. The pooled homogenate and bead rinsings (in isolation medium) are centrifuged for 10 min at 1000 g to remove unbroken cells and debris. Mitochondria are harvested from the supernatant by centrifugation for 10 min at 10,000 g, and washed by resuspension and recentrifugation. All manipulations are performed at 4°C. Mitochondrial respiration is measured in the same oxygen electrode assembly as for whole cells, using 1–5 mg mitochondrial protein per ml isolation medium in an electrode chamber of approximately 2 ml volume at 30°C.

Fig. 5 shows the effect of the tricyclic antidepressant drug chlorimipramine on the respiration of isolated yeast mitochondria. The drug appears to have a biphasic action, uncoupling (and hence stimulating respiration) at low concentrations, while inhibiting respiration at higher concentrations (Linstead, D. & Evans, I. H., unpublished observations).

(3) *Effects on mitochondrial biogenesis*

S. cerevisiae mtDNA is known to encode (Evans 1983): 24 tRNAs, 2 rRNAs (15S and 21S), a mitoribosome associated protein (var 1), 3 subunits of the cytochrome oxidase complex (I, II and III), the apoprotein of cytochrome b, and 2 subunits of the ATPase complex (6 and 8). This can be compared with 22 tRNAs, 2 rRNAs (12S and 16S), 3 subunits of the cytochrome oxidase complex (I, II and III), and the

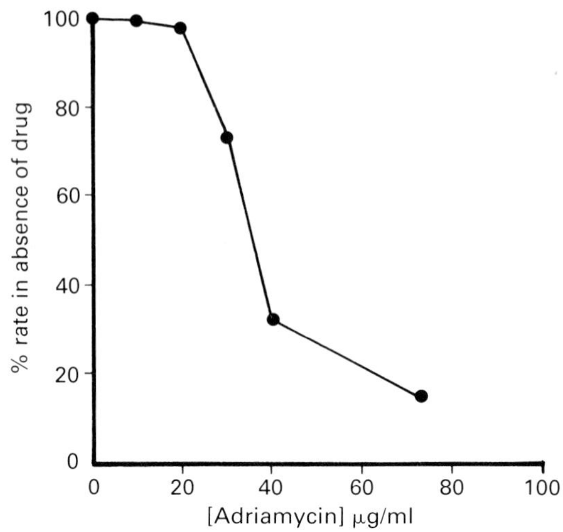

FIG. 4(a). Cells of strain A285 grown in YEG medium were introduced into an oxygen electrode chamber (3 ml volume) containing distilled water at 30°C. The yeast concentration in the chamber was 3.7 mg dry wt/ml (*c.* 2×10^7 cells/ml) and the control (100%) respiration rate was 15.9 nmol oxygen/min. Successive additions of adriamycin were made to the chamber by microsyringe from a 10 mg/ml stock solution in distilled water.

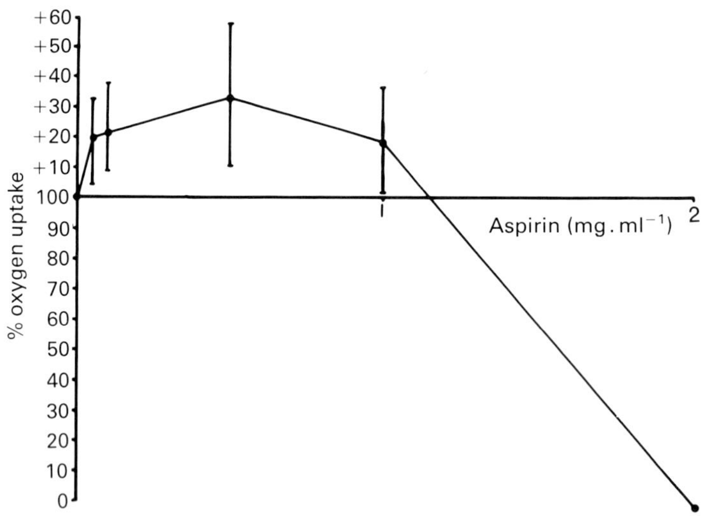

FIG. 4(b). Effect of aspirin on the rate of respiration in yeast cells. Aliquots of approx. 1.5×10^7 cells were taken from an overnight culture of strain B/B in YED medium and inoculated into shake flasks containing 15 ml each of YEG medium and aspirin at concentrations of 0, 0.05, 0.1, 0.5 and 1.0 mg/ml respectively. The flasks were shaken at 30°C for 26 h and cells from each flask were then pelleted, washed twice and suspended in HEPES buffer, pH 6.5. 2 ml of each suspension was added in turn to the chamber of a Clark-type electrode and respiration rate recorded polarographically. Results plotted represent oxygen uptake per 10^7 cells for each culture, and are the average of six experiments. Horizontal line: respiratory rate of control cells.

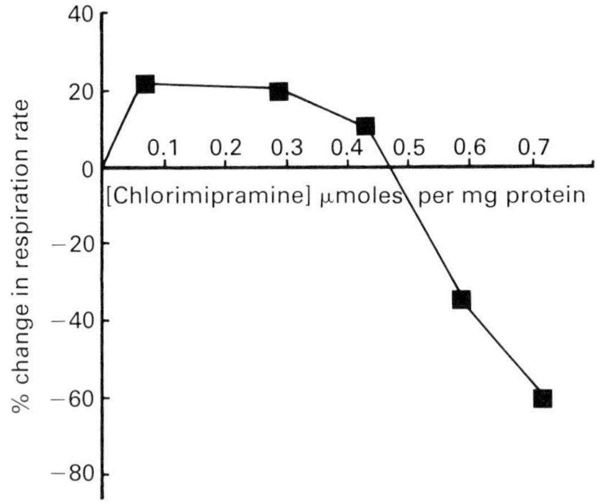

FIG. 5. Mitochondria isolated from strain D6 were introduced into a 2.2 ml electrode chamber containing isolation medium at 30°C. The mitochondrial protein concentration in the chamber was 0.94 mg/ml and the rate of respiration at zero drug concentration in the presence of pyruvate (2.3 mM) and malate (45 μM) was 12 natoms oxygen/min. Successive additions of chlorimipramine to the chamber were made by microsyringe from an aqueous stock solution.

apoprotein of cytochrome b (plus some unassigned open reading frame) encoded in mammalian mtDNA.

Key inner membrane components are specified and fabricated within the mitochondrion itself. Toxicant-induced damage to the mitochondrial genetic system, either through mutagenesis or interference with mtDNA replication, transcription or translation could bring about an impairment of the organelle's respiratory competence. Yeast's ability to continue growing in the presence of a fermentable carbon source when mitochondrial respiration is defective enables toxicant effects on mitochondrial biogenesis to be observed *in vivo*. It has permitted the isolation of a wide range of mutants which have greatly advanced the understanding of the (yeast) organelle's molecular biology. It is possible, for example, to assess directly effects on organelle transcription and translation *in vitro* by measuring incorporation of radiolabelled substrate into mitochondrial RNA or protein (Evans *et al.* 1973; Murray & Linnane 1972). However, yeast provides two very simple *in vivo* approaches for the detection of primary effects on mitochondrial biogenesis.

(i) *Studies on whole cell cytochrome content*

Yeast cells (of a suitable sensitive strain) are grown to stationary phase in liquid YED medium in the presence of toxicant at a concentration inhibitory to growth on YEG. The cells are harvested by centrifugation (5 min at 7000 rpm in a bench top centrifuge) and washed twice in distilled water (or buffer) and resuspended in water or buffer as a thick cream (approximately 1 g of cells/ml of liquid). Cells in this condition are (usually) in a reduced state (sodium dithionite can be added to a concentration of a few μg/ml to ensure this) with the mitochondrial cytochromes exhibiting a very characteristic absorbance profile in the visible spectrum (Fig. 6). The cytochrome absorption spectrum can be readily detected with a simple spectrophotometer such as the Pye–Unicam SP1800 double beam spectrophotometer. A blank of several thicknesses of filter paper is used to zero the machine. Typical inhibitors of mitochondrial biogenesis (such as chloramphenicol or erythromycin) can partially or completely suppress the absorption peaks at 562 nm and 603 nm produced by cytochrome b and cytochrome oxidase respectively. However, there is little if any effect on the peak at 552 nm produced by cytochrome c. Effects of some compounds on cytochrome spectra of yeast cells are illustrated in Fig. 6. Cytoplasmic *petite* mutants of yeast have a very similar absorbance spectrum

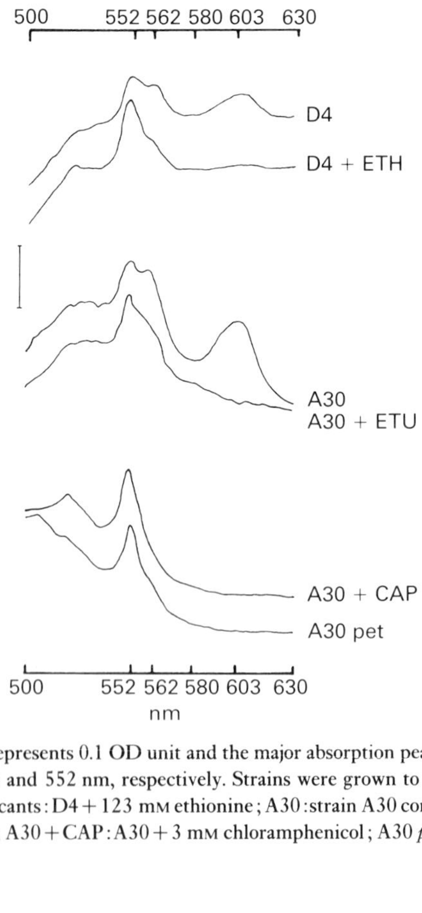

FIG. 6. The vertical bar represents 0.1 OD unit and the major absorption peaks for cytochromes aa_3, b, and c are 603 nm, 562 nm and 552 nm, respectively. Strains were grown to stationary phase in liquid YED with and without toxicants: D4 + 123 mM ethionine; A30: strain A30 control; A30 + ETH: A30 + 100 mM ethylene thiourea; A30 + CAP: A30 + 3 mM chloramphenicol; A30 p^-: *petite* mutant of A30.

to cells exposed to inhibitors of mitochondrial transcription or translation (Fig. 6). Spectral resolution can be imporoved by low temperature spectroscopy (Luzikov *et al.* 1971).

(ii) *Effects of toxicants on aerobic adaptation of anaerobic cells*

Yeast cells grown anaerobically on glucose contain mitochondria which are incapable of oxidative phosphorylation. They appear morphologically simpler than those of cells grown in glycerol and are less biochemically developed. They are often referred to as promitochondria (Wallace *et al.* 1968). On exposure to oxygen and in the absence of glucose repression there is a surge in mitochondrial transcription and translation, culminating in the formation of respiratory competent mitochondria. Toxicants which interfere with mitochondrial transcription and translation tend to suppress this process.

S. cerevisiae grows adequately on a rich glucose medium with appropriate supplements, when in a strictly anaerobic environment, for example in an anaerobic jar flooded with nitrogen (Asano, 1972; Djavadi-Ohaniance *et al.* 1978).

Respiratory adaptation is monitored by measuring respiratory activity of samples of cell suspensions in the oxygen electrode (see previous section). It is also useful to monitor cell viability and *petite* frequency at the end of the adaptation period. Table 2 illustrates the effect on respiratory adaptation of two drugs,

TABLE 2. *The effects of adriamycin and acriflavine on the aerobic adaptation of anaerobically grown yeast (strain A285)*

		Adriamycin				Acriflavine			
Drug concentration (μg/ml):	0	50	100	200	300	5	10	20	50
Growth as % control	100	82	83	63	2	72	68	62	67
Respiration (nmol O_2/min/mg dry wt)	39.8	24.7	17.2	11.3	0.6	0.0	—	—	—
Respiration as % control	100	62	43	28	1.5	0	—	—	—
% *petite*	10	7	8	7	—	89	99	100	100
% viable	100	18	26	25	<0.01	100	76	*c.* 50	*c.* 50

Strain A285 was pregrown to stationary phase at 30°C in N_2-flushed anaerobic jars on supplemented glucose medium (Asano 1972). The cell sediments were resuspended in 'adaptation medium' (1.5% yeast extract, 3% glucose) and distributed between 100 ml conical flasks with side arms (10 ml cell suspension per flask). Flasks, containing different concentrations of adriamycin or acriflavine, were incubated aerobically at 30°C. Respiratory rates were measured after 44 h aerobic adaptation and dry weight concentrations were determined from samples at 44 h and 48 h (end of experiment). At 48 h samples of cell suspension were counted (haemacytometer) and plated on YED for viability determination, and on PDM (1% yeast extract, 0.2% glucose, 4% glycerol) for determining *petite* frequencies from numbers of small colonies.

adriamycin and acriflavine. Acriflavine is a DNA-intercalating dye which selectively inhibits mtDNA replication and transcription. Even at the lowest concentration tested (5 μg/ml) acriflavine totally suppresses respiratory adaptation, while not significantly affecting viability. By contrast, adriamycin (a DNA-binding anthracycline antibiotic used in cancer chemotherapy) is much more toxic in its effect on cell viability than inhibiting in respiratory adaptation.

An important consideration in this type of experiment is that the toxicant is acting initially on anaerobically grown cells, and that activation mechanisms found in aerobic cells may not be operative (see Case Study 1).

4. *Mitochondrial Mutagenesis*

A number of chemically disparate mutagens are now known to interact selectively with or mutagenise the mtDNA of higher eukaryotes and yeast (Wilkie & Evans 1982; Wilkie *et al.* 1983). *S. cerevisiae* is particularly suitable for observing the induction of mitochondrial mutants because the organism is capable of surviving and growing after having lost part or whole of the functionality of the mitochondrial genome. Also many of the mutations occur at a high frequency and are easily scored or selected. Nuclear mutation or chromosomal rearrangement can also be studied in the same strain so that parallel observations of mutagenicity to both nucleus and cytoplasm can be made.

(i) *The* petite *mutation*

This non-reverting deletion mutation (Bernardi 1979) was the first cytoplasmic mutation to be studied in yeast: the mtDNAs of many *petite* mutants have now been sequenced, and it has been observed that massive losses of wild type (*wt*) sequences can occur, together with reiteration of a retained sequence which must include an origin of replication (Bernardi 1979). The details of the molecular mechanisms involved are unclear but recombinational events within the mtDNA are probably important. *Petite* mutations occur at a high spontaneous frequency in many *S. cerevisiae* strains (generally at about 1%) and *petite* colonies have a characteristic phenotype on glucose medium: they appear white when compared with the creamy-coloured colonies produced by *wt* cells. Replica plating (Lederberg & Lederberg 1952) or toothpicking onto solid YEG medium of suspected *petite* colonies confirms inability to grow on a non-fermentable medium. Respiration deficiency of *petite* colonies can also be determined using the tetrazolium overlay technique in which a top agar containing a tetrazolium redox dye is poured over a YED plate bearing mature *wt* and putative *petite* colonies in late log phase: actively respiring colonies reduce the dye and become a deep red, *petite* colonies remain white (Nagal 1959).

High spontaneous and induced frequencies of heritable respiratory deficiency in yeast are a good indication of the cytoplasmic nature of the mutation which can be demonstrated genetically by the appropriate crosses (for a general survey of yeast genetic methods see Sherman *et al.* 1983). A cross between a haploid *petite* and a haploid *wt* of appropriate mating type produces a stable diploid which is usually *wt*. Sporulation of the diploid will yield meiotic tetrads exhibiting the non-Mendelian ratio of 4 *wt*:0 *petite*. Some *petites* may be suppressive, which means that zygotes from a *petite* × *wt* cross yield diploids in which mitotic segregation of *petites* occurs at a high frequency. Diploids which are *petite* cannot sporulate.

Petite frequencies can be scored directly from plating on YED of cultures exposed to toxicants. Selection of *petites*, as opposed to induction, can be assessed by comparing toxicant resistances of a sample of the putatively induced mutants and the *wt* parent strain.

As Table 2 shows for adriamycin and acriflavine, while some anti-mitochondrial agents can have virtually no effect on the *petite* frequency, others can induce *petites* with close to 100% efficiency.

(ii) *Point mutations*

A great variety of point mutations are now known, affecting virtually all known gene products of the yeast mtDNA (Evans 1983). Some, located in the rRNA genes or ATPase subunit genes, confer resistance to antibiotics such as chloramphenicol, erythromycin and oligomycin, and are readily scored. The growth of most haploid yeast strains on YEG is inhibited at concentrations of: (a) chloramphenicol, 0.2–1.0 mg/ml, (b) erythromycin, 1–4 mg/ml, and (c) oligomycin, 0.5–1.5 μg/ml, while mitochondrial resistance mutations can confer full resistance to these antibiotic concentrations. The other major class of mitochondrial point mutation occurs in genes such as those specifying cytochrome b and cytochrome oxidase subunits which produce cellular phenotypes identical to those of *petites*, i.e. inability to grow on non-fermentable medium. However, mitochondrial protein synthesis still occurs, which is almost always totally absent in *petites*. These mutants are classes *mit*$^-$, and both *mit*$^-$ and antibiotic drug resistant mutants show non-Mendelian inheritance, recombining and segregating freely in zygote clones (Wilkie & Thomas 1973).

Table 3(a) shows apparent induction of high frequencies of antibiotic resistant mutants after treatment with the drug adriamycin, and Table 3(b) shows that of the mutants tested in genetic crosses, all exhibited segregation, indicating that the mutations are indeed located in mtDNA.

Case Study 1: Adriamycin

Adriamycin is an anthracycline antibiotic in current chemotherapeutic use against a wide range of tumours.

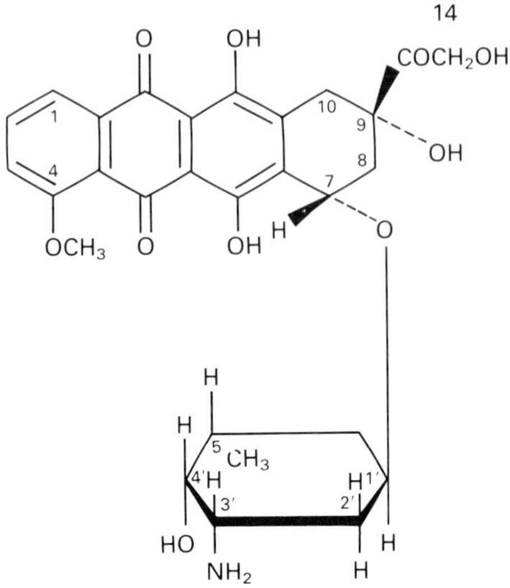

It has a number of toxic side effects, cardiomyopathy probably being the most serious, which limit its clinical exploitation. Adriamycin is able to bind to DNA by intercalation, inhibits many DNA-dependent functions, and induces mutations and neoplasia. While adriamycin can certainly bind to nuclear DNA and disrupt chromatin organization, the drug is also able to damage other cellular organelles, including the plasma membrane and the mitochondrion. Gianni and colleagues have written a recent general review of anthracycline toxicity and anti-tumour activity (Gianni *et al.* 1983).

In view of the complexity of the drug's interactions with mammalian cells, it was of interest to test it in the yeast system.

Preliminary screening on solid medium indicated *primary* antimitochondrial activity in a number of strains (Egilsson, V. & Evans, I. H., unpublished results), one strain in particular, A285, showing a very marked contrast in sensitivities on glucose and glycerol media. Growth tests in liquid medium with selected strains confirmed this finding, the results for A285 being depicted in Fig. 3(a). Oxygen electrode studies show that the drug can certainly inhibit the respiration of whole yeast cells (see Fig. 4(a)). The respiration of isolated yeast mitochondria is also sensitive to adriamycin (see Table 4), though much higher concentrations are required to significantly inhibit respiration than are needed to block growth on glycerol medium. These results show that while adriamycin can directly disrupt functions of the inner membrane, the drug's potency is lower than many classical respiratory inhibitors, and other mechanisms of inhibitory action on yeast cell growth on non-fermentable medium should be considered.

The ability of anthracyclines to intercalate into DNA and block various DNA-

TABLE 3(a). *Induction of antibiotic-resistant mutants by adriamycin*

Length of treatment (35 μg/ml ADR)	Total cells plated on: ER	OL	Viability (%)	No. of resistant colonies: ER	OL
CONTROL	2×10^6	2×10^6	100	15 (7.5)*	12 (6)
15′	2×10^6	2×10^6	20	11 (27)	58 (145)
30′	10^6	7.5×10^6	14	14 (100)	25 (240)
60′	2×10^5	2×10^5	7	8 (570)	17 (1380)
120′	2×10^5	2×10^5	3	21 (3500)	52 (8650)

Strain Ber B was pregrown in YED and resuspended in distilled water containing 35 μg/ml adriamycin, at approximately 5×10^7 cells/ml. The cells were incubated statically at 30°C. Samples were taken at intervals, the cells washed in distilled water and plated on YEG containing erythromycin (ER) or oligomycin (OL), and on YED, to determine viability.

* Figures in parentheses are numbers of resistant colonies per 10^6 viable cells.

TABLE 3(b). *Genetic analysis of adriamycin-induced, erythromycin- and oligomycin-resistant mutants*

		R	S			R	S
Ber B ER^R mutant no.	1	11	3	Ber B OL^R mutant no.	1	11	4
	2	8	5		2	17	10
	3	3	1		3	11	4
	4	4	8		4	6	6
	5	13	5		5	8	6
	6	12	6		6	11	2

Mitotic segregations in the diploid progeny of drug resistant mutants × wild type (sensitive) crosses. Independently isolated drug-resistant mutants isolated from erythromycin (ER) or oligomycin (OL) YEG plates were crossed to a drug-sensitive tester strain (Ber A). Zygote clones were sampled and plated on YED and approximately fifteen randomly selected diploid clones were treated for drug resistance. Mitotic segregation of drug resistance (R) versus sensitivity (S) indicates cytoplasmic (mitochondrial) inheritance.

dependent functions suggests that adriamycin might be an inhibitor of mitochondrial biogenesis, and, indeed, adriamycin's congener, daunomycin, is able to inhibit mitochondrial transcription *in vitro* (Evans *et al.* 1973). However, aerobic adaptation experiments yielded unexpected results. As Table 1 shows, adriamycin is far less efficient at suppressing aerobic adaptation than acriflavine, a well-characterized inhibitor of mitochondrial transcription. It can also be seen from Table 2 that exposure of adapting cells to adriamycin has no significant impact on the *petite* frequency, whereas acriflavine is a powerful *petite* inducer.

The results so far suggest that while adriamycin is a strongly selective inhibitor

TABLE 4. *Effects of adriamycin on the respiration* of mitochondria isolated from strain A285*

Adriamycin conc. (μg/ml)	Rate respiration (nmol O_2/min)	% inhibition
7	25.0	0
32	25.0	0
82	26.3	0
132	23.6	5
182	19.3	23
232	13.1	47.5
282	11.8	53
332	11.4	54
+KCN	0.9	96

The respiration of mitochondria isolated from strain A285 was assessed under the conditions described for Fig. 5, except that ADP was not rate-limiting. Successive additions of adriamycin were made from a 10 mg/ml aqueous stock to give final electrode concentrations as indicated. 50 mM potassium cyanide (KCN) effectively inhibited respiration.

* Respiration is state 3 (ADP stimulated) with pyruvate and malate as substrates.

of growth on non-fermentable medium, it is not acting like conventional inhibitors of mitochondrial function or biogenesis. The low activity of adriamycin against initially anaerobic cells observed in the adaptation experiments could indicate that oxygen is an important factor in adriamycin toxicity. Support for this is found in Table 5, where it is clear that oxygen powerfully enhances adriamycin's lethality and the frequency of *petite* mutants in the culture. There is evidence that *petite* mutants can be more resistant to oxygen-potentiated adriamycin toxicity (killing) than the respiratory competent parent strain (Hixon, S., personal communication) and this could imply a mitochondrial role in the adriamycin-mediated generation of oxy-radicals. Work on mammalian systems has shown that microsomal NADPH cytochrome p450 reductase can transfer electrons from NADPH to adriamycin, a quinone, reducing it to a semiquinone free radical which in turn can interact with oxygen to form oxy-radicals (Bachur *et al.* 1979); a similar mechanism involving enzymic reduction of adriamycin may be responsible for the ability of mammalian submitochondrial particles to generate superoxide on interaction with adriamycin (Thayer 1977). Adriamycin and oxy-radicals might be expected to cause both intra- and extra-mitochondrial damage. The adriamycin species, with its DNA-binding propensity, is presumably more site-specific. As expected on this view of adriamycin toxicity, free radical scavengers can antagonize the drug's toxicity in yeast (Nudd & Wilkie 1983) and mammalian systems (Myers *et al.* 1976).

Increased *petite* frequencies in adriamycin-treated yeast cultures might well be

TABLE 5. *Effect of anaerobiosis on adriamycin-toxicity and* petite *induction in strain D6*

Inoculum	Culture	Adriamycin conc. (μg/ml)	Viability (%)	*Petite* Frequency (%)
Aerobic	Anaerobic	0	100	2
		50	47	2
		75	47	14
		100	20	14
Anaerobic	Anaerobic	0	100	4
		50	73	6
		75	73	8
		100	25	8
Aerobic	Aerobic	0	100	1
		50	9	25
		75	12	37
		100	2	61
Anaerobic	Aerobic	0	100	1
		50	10	?
		75	6	82
		100	2	62

Strain D6 was pregrown aerobically in YED or anaerobically in supplemented glucose (Asano 1972) medium. The cultures were used to inoculate flasks containing liquid YED or Asano medium and adriamycin at various concentrations. These cultures were incubated at 30°C for approximately 16 h statically (anaerobic) or with vigorous shaking in air (aerobic), then sampled and plated on YED for viability estimation and on PDM (see Table 2) for determination of *petite* frequencies.

attributable to selection rather than induction, given the probable role of respiratory activity in adriamycin toxicity. However, Table 3(a) presents evidence that extensive induction of mitochondrial point mutations can occur: high frequencies of erythromycin- and oligomycin-resistant mutants were found in an adriamycin-treated yeast strain and all the mutants analysed genetically showed mitotic segregation indicative of cytoplasmic inheritance (Table 3(b)).

In summary, a number of simple tests have shown that mitochondria are a key target in the cytotoxic action of adriamycin against yeast. It appears that adriamycin interacts with respiring yeast mitochondria to generate a variety of highly destructive free radical species, including adriamycin- and oxy-based species. These attack various cell targets inside and outside the mitochondrion, causing both mutogenesis of mitochondrial DNA and cell death. The yeast findings are consistent with information from mammalian systems, and suggest that the role of mammalian mitochondria in adriamycin intoxication of mammalian tissues should be closely examined.

Case Study 2: Aspirin

Aspirin is a non steroidal anti-inflammatory drug which is known to affect a wide range of metabolic activities in mammalian systems (Smith & Dawkins 1971). Two of the drug's documented effects are that it is able to uncouple oxidative phosphorylation and depress mitochondrial ATP synthesis (Brody 1956; Smith & Jeffrey 1956). Since aspirin is of such great pharmacological interest, it seemed useful to study its specific effects on the yeast mitochondrial system.

Initial experiments on solid YED and YEG media using an aspirin concentration range of 0–3.5 mg/ml indicated that at concentrations of between 1.5 and 2.5 mg/ml aspirin was inhibitory to growth on YEG but not YED. The MIC of aspirin on YED medium for the strains tested ranged between 2.5 and 3.5 mg/ml. The largest difference between MICs on YED and YEG for a specific strain was 3-fold (Table 1). These results suggested that aspirin was more inhibitory to the mitochondrial system than other cellular processes at high concentrations.

A more detailed assessment of aspirin's effect on the growth of strains Ber B and A30 in liquid YED and YEG media (Fig. 3(b)) indicated that the drug's effect was biphasic. At lower concentrations, 0.05–1.0 mg/ml, it stimulated growth in YEG, whilst at concentrations greater than this aspirin selectively inhibited growth in YEG. The MIC in YEG was 2.0 mg/ml. In YED medium no stimulation of growth in the respiratory phase was observed at concentrations which potentiated growth on YEG but at 2.0 mg/ml a selective inhibition of the respiratory phase of growth was observed. The MIC in YED was 3 mg/ml. This evidence confirmed the previous data generated from solid media tests that aspirin selectively inhibits yeast cell mitochondria at high concentrations and further implies that at low concentrations aspirin has a stimulatory effect on the organelle.

The effect of aspirin on the respiration rate of whole yeast cells of strain Ber B grown in YEG medium containing the drug was also tested. Concentrations of aspirin used were 0.05, 0.1, 0.5, 1.0 and 2.0 mg/ml and results are illustrated in Fig. 4(b). It was clear that at concentrations between 0.05 and 1.0 mg/ml aspirin stimulated oxygen uptake by yeast cells, and maximum uptake rate was observed in the culture treated with 0.5 mg/ml. This correlated well with the increased growth rate previously observed in YEG medium at these aspirin concentrations. Conversely, the complete inhibition of respiration observed at 2.0 mg/ml correlated well with the lack of a respiratory phase of growth at this aspirin concentration observed in YED medium.

The mutagenic properties of aspirin, with respect to mtDNA, were studied by testing its ability to induce the *petite* mutation in a number of yeast strains. It was observed that at 2.0 mg/ml aspirin induced the *petite* mutation in all strains tested. The degree of *petite* induction was strain dependent. The largest factor of induction was eighteen times for strain 188 wt (Table 6). Aspirin also considerably reduced cell viability at 2.0 mg/ml.

To further define aspirin's activity with respect to the mitochondrial system its ability to reverse *petite* induction by the mitochondrial mutagens ethidium bromide (EB) and adriamycin (ADR) was studied. EB and ADR are DNA-intercalating compounds (Lerman 1961; Zunio *et al.* 1972) which induce the *petite* condition in yeast (Slonimski *et al.* 1968; Evans & Wilkie 1980). Mitochondrial mutagenicity was confirmed as shown in Fig. 7. It can also be seen from Fig. 7 that the yeast cell mitochondria were protected from mutagenesis by EB and ADR in a dose-dependent manner at aspirin concentrations between 0.05 and 1 mg/ml. At a concentration of 1.75 mg/ml, aspirin had little or no effect on *petite* induction by these mutagens. This may be a critical concentration of aspirin above which it becomes toxic to the mitochondria and itself induces the *petite* condition (Bruce & Wilkie 1983). Protection against EB mutagenesis was effective but only over a period of up to about 8 h. This was probably due to the fact that the dose used of EB was high (100% *petite* induction effected in 4 h): the aspirin effect may be more pronounced in this system if EB concentrations giving about 50% *petite* were used.

Results shown in Fig. 7 give the findings over a 10 h period, but in subsequent experiments with ADR, cells were sampled after an incubation period of 24 h. After this time, cultures treated with 0.07 mg/ml ADR alone showed a *petite* frequency of

TABLE 6. *Effect of aspirin on the mitochondrial mutation petite (p^-) and viability in various strains of yeast*

Treatment	Strain	No. of colonies	Cell viability* (%)	No. p^-	p^- (%)
Controls	B/B	456	100	3	0.66
	188	1068	100	1	0.09
	D6	801	94	5	0.66
	A30	1084	100	9	0.93
	D18	709	100	4	0.56
Aspirin (2 mg/ml)	B/B	2060	23	58	2.80(4)†
	188	1493	10	23	1.54(18)
	D6	1424	14	45	3.16(5)
	A30	300	4	24	8.0(9)
	D18	1217	19	32	2.63(5)

Samples of strains Ber B, 188 wt, D6, A30 and D18 were removed from control (0 mg/ml aspirin) and 2.0 mg/ml aspirin YED plates of the previous experiment (see Table 1) and suspended in sterile distilled water. The cell density was measured using a haemacytometer and 0.1 ml of a serial dilution of the suspensions plated in triplicate for each strain and aspirin concentration onto YED for viability estimation and PDM for determination of *petite* frequencies (see Table 2).

* Ratio of colonies to cells plated.

† The numbers in parentheses represent × spontaneous rate.

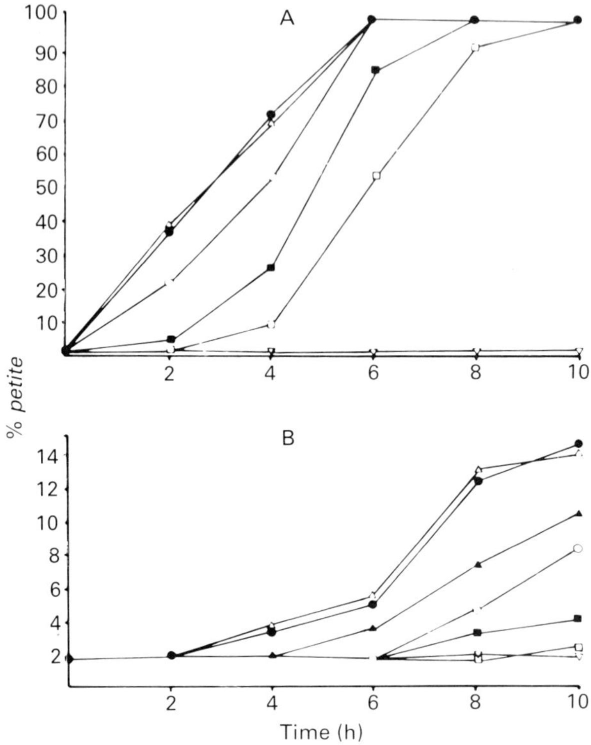

FIG. 7. Protective effect of aspirin against mitochondrial mutagenesis by ethidium bromide and adriamycin. (A) Ethidium bromide (EB). Yeast strain B/B was grown overnight in 1% yeast extract, 2% glucose (YED) medium. Aliquots of 1.5×10^7 cells were inoculated into shake flasks containing YED and EB (20 μg/ml). A control flask was also set up which had no EB. Aspirin (ASP) was added to EB-containing flasks to give concentrations of 0, 0.1, 0.5, 1.0 and 1.75 mg/ml respectively. Flasks were shaken at 30°C. At zero time and thereafter at 2 h intervals, cells were sampled from each flask and plated. *Petite* colonies (mitochondrial deletion mutants) were identified by their small size and inability to grow when transferred from YED on which they were growing, to non-fermentable, yeast-extract-glycerol medium. ▽, YED control; ●, EB; ○, EB+0.1 mg/ml ASP; ■, EB+0.5 mg/ml ASP; □, EB+1.0 mg/ml ASP; △, EB+1.75 mg/ml ASP. (B) Adriamycin (ADR). The same protocol was used, substituting 0.07 mg/ml ADR for EB. Symbols are also the same, with the additional growth curve for ADR+0.05 mg/ml ASP (▲).

about 35%, but in the presence of 1 mg/ml aspirin in the culture medium this frequency was reduced to about the spontaneous frequency (approximately 1%). Lower concentrations of aspirin (0.1–0.5 mg/ml) also reversed *petite* induction by ADR over this time period but proportionately less effectively.

Finally, the effect of aspirin on the antimitochondrial activity of oligomycin (OL) was investigated. OL is not mutagenic but selectively blocks oxidative phosphorylation by binding to a component of the ATPase complex (Tzagoloff 1969). The results of tests on seventeen yeast strains are given in Table 7 from which it is apparent that aspirin reversed the mitochondrial inhibition by OL in most strains. The protective effect was proportional to aspirin concentration and, once again, within the range 0.1–1 mg/ml. At the higher concentration of 1.75 mg/ml, aspirin showed mitochondrial toxicity as in the previous experiments. A more accurate assessment of the effect of aspirin on OL inhibition was obtained from growth curves in liquid medium. The results (Fig. 8) again show a dose-related reversal of the inhibition with the concentration of 1 mg/ml aspirin as the most efficient dose.

To summarize, these results indicated that aspirin has a specific anti-mitochondrial effect which was biphasic. At low concentrations, 0.05–1.0 mg/ml, it stimulated respiration, protected mtDNA from mutagenesis by the mitochondrial mutagens EB and ADR and reversed inhibition of oxidative phosphorylation by oligomycin. Conversely at concentrations above 1 mg/ml aspirin became selectively

TABLE 7. *Inhibition of growth by oligomycin in various yeast strains: reversal by aspirin*

	Minimum inhibitory concentration of oligomycin (μg/ml)				
		With aspirin (mg/ml)			
Strain	Without aspirin	0.1	0.5	1.0	1.75
D11	0.5	1.0	1.0	1.5	<0.25
A30	0.5	0.5	1.0	1.25	<0.25
B21	0.5	0.75	0.75	0.75	<0.25
6/81	0.5	0.5	0.5	0.5	<0.25
B/B	0.5	0.75	0.75	1.0	<0.25
D18	0.25	0.5	0.5	1.0	<0.25
B27	0.5	0.5	0.75	1.0	<0.25
D22	0.25	0.25	0.25	0.5	<0.25
2180a	1.0	<1.5	<1.5	<1.5	<0.25
2180	0.75	0.75	1.5	<1.5	<0.25
D4	0.25	0.5	0.5	0.75	<0.25
B41	0.5	0.5	0.5	0.5	<0.25
A285	1.25	<1.5	<1.5	<1.5	<0.25
B/A	0.25	1.0	1.0	1.25	<0.25
D6	0.25	0.75	0.75	0.75	<0.25
1880LI^{r}*	>1.5	>1.5	>1.5	>1.5	<0.25
41/1610LI^{r}*	>1.5	>1.5	>1.5	>1.5	<0.25

Seventeen *S. cerevisiae* strains were grown to stationary phase on YEG medium and cells from each strain drop inoculated simultaneously onto five series of petri dishes in which the carbon source was glycerol and oligomycin was present at 0, 0.25, 0.5, 0.75, 1.0, 1.25 and 1.5 μg/ml. The plates were incubated at 30°C for 3 days and subsequently scored visually for growth.

* Oligomycin resistant mutants.

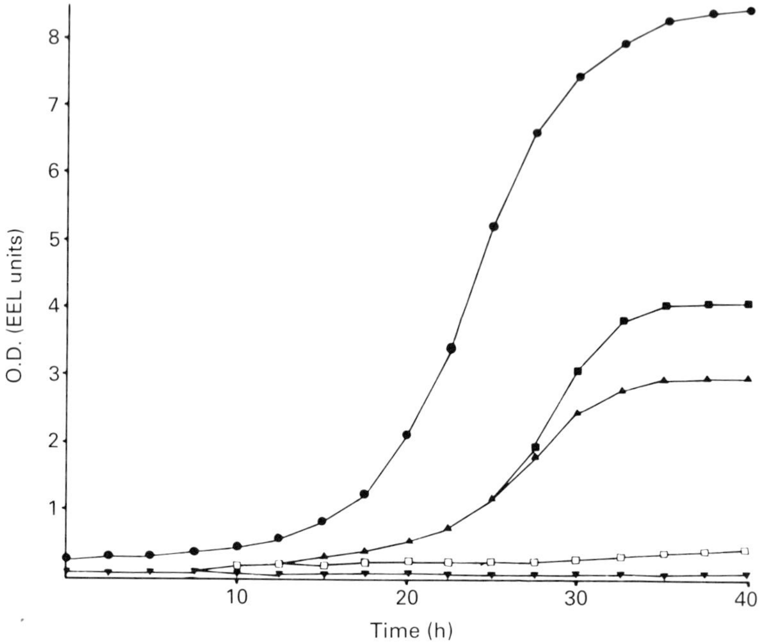

FIG. 8. Inhibition of growth by oligomycin (OL): reversal by aspirin. Aliquots of approx. 1.5×10^7 cells of yeast strain 2180α cultured overnight in YED medium, were inoculated into five flasks containing 15 ml each of liquid YEG (non-fermentable) medium. OL was added to four of the flasks to a concentration of 1.25 μg/ml in each. Aspirin was then added to three of the OL-containing flasks to give ASP concentrations of 0.1, 0.5 and 1.0 mg/ml respectively. Flasks were shaken at 30°C and O.D. readings taken regularly using an EEL colorimeter. ○, YEG control; ▽, OL; ○, OL + 0.1 mg/ml ASP; △, OL + 0.5 mg/ml; □, OL + 1.0 mg/ml ASP.

toxic to the mitochondrial system, inhibiting respiration and inducing the *petite* mutation.

Previous findings from work with aspirin-treated mammalian cells have shown that low concentrations of aspirin non-specifically stimulate cell metabolism, protein turnover and DNA replication (Hial *et al.* 1976). This could be explained by aspirin's primary stimulation of the mitochondrial system at these concentrations, through which general cell metabolism is stimulated.

The data suggest a membrane-associated activity of aspirin (Bruce & Wilkie 1984), possibly indicating an association with the inner mitochondrial membrane at or near the site of the mitochondrial ATPase.

Summary

From these results it is clear that the yeast *S. cerevisiae* performs well as a test organism in determining the anti-mitochondrial activity of compounds, toxicants or otherwise. Its primary benefits as a system are its inexpensive nature, simplicity to maintain, well researched and understood genetics and biochemistry, and speed in generating results. It is attracting interest as a potential screening system for mutagens and carcinogens and has already helped in elucidating the cellular targets of many classes of compounds.

Acknowledgments

We would like to thank our colleagues Mrs Doreen Collier, Dr Valli Egilsson and Dr David Linstead for collaborative work and discussions; and the Cancer Research Campaign, Ciba-Geigy (U.K.) Ltd, and the SERC for financial support. Many thanks also to Mrs Brenda Bates and Mrs Lorraine Lacey for preparation of the manuscript.

References

ASANO, K. 1972 Effect of antibiotics on adaptive formation of mitochondria in yeast. *Journal of Biochemistry* **72**, 737–747.

BACHUR, N. R., GORDON, S. L., GEE, M. V. & KON, H. 1979 NADPH cytochrome P-450 reductase activation of quinone anticancer agents to free radicals. *Proceedings of the National Academy of Sciences of the United States of America* **76**, 954–957.

BERNARDI, G. 1979 The *petite* mutation in yeast. *Trends in Biochemical Science* **4**, 197–201.

BRODY, T. M. 1956 Action of sodium salicylate and related compounds on tissue metabolism *in vitro*. *Journal of Pharmacology and Experimental Therapeutics* **117**, 39–51.

BRUCE, I. J. 1983 Anti-mitochondrial effects of anti-thrombotic drugs. Ph.D. thesis, London University.

BRUCE, I. J. & WILKIE, D. 1983 Selective mitochondrial activity distinguishes aspirin from salicylate and benzoate in yeast cells. *Biochemical Pharmacology* **32**, 2116–2118.

BRUCE, I. J. & WILKIE, D. 1984 Effect of aspirin on mitochondrial mutagens in *S. cerevisiae*. *Molecular and General Genetics* **194**, 299–302.

DJAVADI-OHANIANCE, L., RUDIN, Y. & SCHATZ, G. 1978 Identification of enzymically inactive apocytochrome c peroxidase in anaerobically grown *Saccharomyces cerevisiae*. *Journal of Biological Chemistry* **253**, 4402–4407.

EGILSSON, V., EVANS, I. H. & WILKIE, D. 1979 Toxic and mutagenic effects of carcinogens on the mitochondria of *Saccharomyces cerevisiae*. *Molecular and General Genetics* **174**, 39–46.

ESTABROOK, R. W. 1967 Mitochondrial respiratory control and the polarographic measurement of ADP:O ratios. *Methods in Enzymology* **10**, 41–47.

EVANS, I. H. 1983 Molecular genetic aspects of yeast mitochondria. *Yeast Genetics, Fundamental and Applied Aspects*, eds Spencer, J. F. T., Spencer, D. M. & Smith, A. R. W., pp. 269–370. Springer-Verlag, New York.

EVANS, I., LINSTEAD, D., RHODES, P. M. & WILKIE, D. 1973 Inhibition of RNA synthesis in mitochondria by daunomycin. *Biochimica et Biophysica Acta* **312**, 323–336.

EVANS, I. H. & WILKIE, D. 1980 Effects of adriamycin on the mitochondria of *Saccharomyces cerevisiae*. *Proceedings of the Society for General Microbiology* **6**, 99–100.

Gianni, L., Corden, B. J. & Myers, C. E. 1983 The biochemical basis of anthracycline toxicity and antitumor activity. *Reviews in Biochemical Toxicology* **5**, 1–82.

Herskowitz, I. 1985 Yeast as the universal cell. *Nature* **316**, 678–679.

Hial, V., Horakova, Z., Schaff, R. E. & Beaven, M. A. 1976 Alteration of tumour growth rate by aspirin and indomethacin: studies with two transplantable tumours in mice. *European Journal of Pharmacology* **37**, 367–376.

Lederberg, J. & Lederberg, E. M. 1952 Replica plating and indirect selection of bacterial mutants. *Journal of Bacteriology* **63**, 399–406.

Lerman, L. S. 1961 Structural considerations in the interaction of DNA and acridines. *Journal of Molecular Biology* **3**, 18–31.

Linstead, D. & Wilkie, D 1971 A comparative study of *in vivo* inhibition of mitochondrial function in *Saccharomyces cerevisiae* by tricyclic and other centrally-acting drugs. *Biochemical Pharmacology* **20**, 839–846.

Luzikov, V. N., Zubatov, A. S., Rainini, E. I. & Bakeyeva, L. E. 1971 Degradation and restoration of mitochondria upon deaeration and subsequent aeration of aerobically grown *Saccharomyces cerevisiae* cells. *Biochimica et Biophysica Acta* **245**, 321–334.

Murray, D. R. & Linnane, A. W. 1972 Synthesis of proteolipid protein by yeast mitochondria. *Biochemical and Biophysical Research Communications* **49**, 855–862.

Myers, C. E., McGuire, W. P. & Young, R. C. 1976 Adriamycin: amelioration of toxicity by α-tocopherol. *Cancer Treatment Reports* **60**, 961–962.

Nagal, S. 1959 Induction of the respiration deficient mutation in yeast by various synthetic dyes. *Science* **130**, 1188–1189.

Sherman, F., Fink, G. R. & Lawrence, C. W., revised by Sherman F., Fink, G. R. & Hicks, J. 1983 *Methods in Yeast Genetics.* Cold Spring Harbor Laboratory, P.O. Box 100, Cold Spring Harbor, N.Y. 11724.

Slonimski, P. P., Perrodin, G. & Croft, J. H. 1968 Ethidium bromide induced mutation of yeast mitochondria: complete transformation of cells to respiratory deficient nonchromosomal *petites. Biochemical Research Communications* **30**, 232–239.

Smith, M. J. H. & Dawkins, P. D. 1971 Review: Salicylate and enzymes. *Journal of Pharmaceutical Pharmacology* **23**, 729–744.

Smith, M. J. H. & Jeffrey, S. W. 1956 The effects of salicylate on creatine phosphate and adenosine 3 phosphate in the isolated rat diaphragm. *Biochemical Journal* **64**, 589–592.

Spencer, C., Symons, S. A. & Brant, R. V. 1971 The preparation and some properties of yeast mitochondria. *Archiv für Mikrobiologie* **75**, 246–259.

Strathern, J. N., Jones, E. W. & Broach, J. R. 1981 *The Molecular Biology of the Yeast Saccharomyces*, Vol. 1. *Life Cycle and Inheritance.* 1982. Vol. 2. *Metabolism and Gene Expression.* Cold Spring Harbor Laboratory.

Struhl, K. 1983 The new yeast genetics. *Nature* **305**, 391–397.

Thayer, W. S. 1977 Adriamycin-stimulated superoxide formation in submitochondrial particles. *Chemical–Biological Interactions* **19**, 265–278.

Venitt, S. & Parry, J. M. 1985 *Mutagenicity Testing, a Practical Approach.* IRL Press, Oxford.

Wallace, P. G., Huang, M. & Linnane, A. W. 1968 The biogenesis of mitochondria. II. The influence of medium composition on the cytology of anaerobically grown *Saccharomyces cerevisiae. Journal of Cell Biology* **37**, 207–220.

Wilkie, D. 1972 The yeast cell in antimitochondrial activity of drugs. *Medical and Biological Illustration* **22**, 119–124.

Wilkie, D. & Evans, I. 1982 Mitochondria and the yeast cell surface: implications for carcinogenesis. *Trends in Biochemical Science* **7**, 147–151.

Wilkie, D., Evans, I. H., Egilsson, V., Diala, E. S. & Collier, D. 1983 Mitochondria, cell surface, and carcinogenesis. *International Review of Cytology* Suppl. **15**, 157–189.

WILKIE, D. & THOMAS, D. Y. 1973 Mitochondrial genetic analysis by zygote cell lineages in *Saccharomyces cerevisiae*. *Genetics* **73**, 367–377.

WOODS, L. F. J. & WISEMAN, A. 1980 Benzo(a)pyrene hydroxylase from *Saccharomyces cerevisiae* substrate binding, spectral and kinetic data. *Biochimica et Biophysica Acta* **613**, 52–61.

ZUNIO, F. R., GAMBETTA, A., DIMARCO, A. & ZAUARA, A. 1972 Interaction of daunomycin and its derivatives with DNA. *Biochimica et Biophysica Acta* **277**, 489–498.

Testing the Industrial Environment for Carcinogens/ Mutagens using *Schizosaccharomyces pombe*

PAMELA MCATHEY AND BHARTI PATEL
Department of Food and Biological Sciences, Polytechnic of North London, Holloway, London N7 8DB

The last 25 years have seen increasing public concern about chemicals present in the human environment. Numerous syndromes from birth defects to pre-senile dementia and from childhood hyperactivity to general malaise have been attributed to environmental chemicals, but it is their potential carcinogenicity which has aroused most disquiet. Approximately one-fifth of deaths in industrialized societies are caused by cancer. While some cancers (e.g. breast cancer) are known to have a strong genetical component, epidemiological data from workers in certain specialized occupations and from migrant groups indicate that environmental factors are highly significant in the causation of many cancers (Cairns 1978).

According to the somatic mutation theory of cancer induction, first proposed by Bauer (1928), DNA damage in somatic cells may result in the production of cancerous cells if mutations occur in those genes which control cell multiplication during normal development. The identification of those chemicals present in the environment which are capable of causing such damage to DNA is becoming an increasingly important medical, social and political issue. It has been estimated that in excess of 50 000 synthetic chemicals are currently in commercial use and up to 1000 additional chemicals are marketed annually (Ames 1979). Only a small proportion of these chemicals has been adequately tested for genotoxicity. Traditionally the identification of environmental carcinogens relied upon epidemiological surveys and long-term animal experiments. However, such tests are of necessity both time-consuming and expensive. In the last 10 years a high degree of correlation between mutagenicity and carcinogenicity has been established and this relationship exploited in the development of several systems, notably the Ames test (McCann *et al.* 1975), designed to detect mutagenic activity and hence predict carcinogenic activity. At present, human risk assessment is made on the basis of a battery of such *in vitro* tests supplemented by appropriate *in vivo* animal experiments

Industrial Microbiological Testing
0-632-01793-7

and epidemiological data. However, the current *in vitro* systems have two major disadvantages:

(1) In order to be both quick and economic, they rely mainly on the use of bacteria such as *Salmonella typhimurium* and *Escherichia coli* as test organisms. The validity of extrapolating from data obtained with bacterial prokaryotic systems for the prediction of cancer induction in human eukaryotic cells is open to question.

(2) The test organism is subjected to an acute dose of the chemical (high concentration/short exposure), but environmentally induced cancers are more likely to be caused by chronic exposures (low concentration/long exposure).

We have developed a system for measuring chemically induced mutation frequencies in continuous cultures of the fission yeast, *Schizosaccharomyces pombe*. *Schizosaccharomyces pombe* shares many of the advantages of bacterial systems (e.g. ease of handling, speed and economy of data collection), but has the additional benefit of being a eukaryotic organism and hence having a cell structure much more closely related to human cells than are those of bacteria. Furthermore, as this system employs continuous cultures, the effect of chronic exposures to environmental mutagens can be assessed.

Mutation Studies in Continuous Cultures

The basic concepts of continuous culture were simultaneously formulated in 1950 by Novick & Szilard and by Monod. If fresh medium is continually supplied to a growing culture of a unicellular organism, the cell population will reach a steady state of growth and division in which the total number of cells and the rate of cell division remain constant with time. To maintain this steady state the volume of the culture must be kept constant, viz. spent medium must be removed from the culture vessel at the same rate at which fresh medium is supplied. In chemostats growth rates are established independently of cell concentration, being solely dependent upon the rate of addition of fresh growth medium. Therefore, at steady state, $\mu = D$, where μ is the specific growth rate and D the dilution rate. The application of continuous culture techniques to mutation studies is discussed at length by Kubitschek (1970). A forward mutation system has to fulfil three main criteria before it can be used for mutational studies on continuous cultures:

(1) The mutants obtained must be amenable to effective screening.

(2) The spontaneous and/or induced mutation frequency must be sufficiently high to permit the recovery of mutants at the cell concentrations which can be maintained in the chemostat.

(3) Mutant cells must not have a selective growth advantage or disadvantage over the non-mutant cells. Fig. 1 shows the increase in mutant numbers with time obtained in three idealized situations: (a) mutant cells under positive selection, (b) no selection, and (c) mutant cells under negative selection. Only in the second

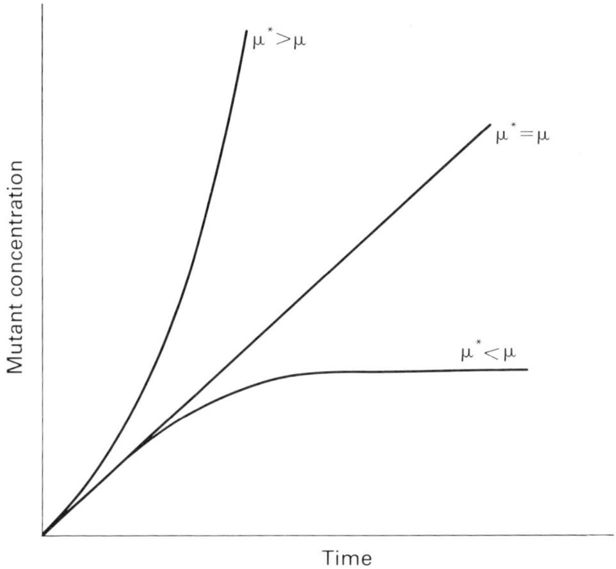

FIG. 1. Theoretical patterns of mutant accumulation in continuous culture. μ^*, growth rate of mutant cells; μ, growth rate of non-mutant cells.

case is a linear increase in mutant numbers with time observed, permitting accurate determination of mutation rates.

Use of *Sch. pombe* as a test organism

The physiology, biochemistry and genetics of *Sch. pombe* have been extensively reviewed (Mitchison 1970; Gutz *et al.* 1974; Egel *et al.* 1980; Phipps *et al.* 1985). Although this yeast has not been utilized as an experimental organism to the same extent as the budding yeast, *Saccharomyces cerevisiae*, comparative studies have shown it to be as sensitive to chemical mutagens as *S. cerevisiae* (Loprieno *et al.* 1983) and it has the additional advantage of possessing only three linkage groups (Gygax & Thuriaux 1984), thereby facilitating genetical analysis.

The kinetics of spontaneous mutation accumulation in chemostat-grown *Sch. pombe* are dependent on the growth-limiting nutrient. Under conditions of glucose-limitation, mutation accumulation is directly proportional to the rate of cell division, while under amino acid-limitations, in appropriate auxotrophic strains, mutation accumulation becomes proportional to chronological time (McAthey & Kilbey 1977; McAthey & Kilbey 1978). It is therefore necessary to monitor mutation frequencies under standard conditions of cell generation time and concentration of growth limiting nutrient.

As it has been suggested that certain carcinogens exhibit greater mutagenic activity on mitochondrial DNA than on nuclear DNA (Egilsson *et al.* 1979) we feel it is important to monitor the effect potential environmental mutagens have on mitochondrial DNA (as judged by any increase in the yield of chloramphenicol and erythromycin-resitant mutations above the respective spontaneous levels) as well as their effect on nuclear DNA (as judged by any increase in the yield of trichodermin and cycloheximide-resistant mutations above the respective spontaneous levels). Forward mutation to resistance to these four antibiotics satisfies the criteria outlined in the previous section and hence they can be usefully employed in mutation studies on continuous cultures of *Sch. pombe* (McAthey & Kilbey 1976; McAthey & Patel, unpublished data). The molecular basis of resistance to these antibiotics is discussed by Gale *et al.* (1981).

Apparatus and Materials

Chemostat

A chemostat essentially consists of four parts, a growth chamber, a nutrient supply system, a gaseous supply system and an overflow system (Fig. 2). Chemostats are commercially available, but perfectly adequate chemostats can be constructed from standard laboratory glassware (see, for example, Kubitschek 1970; Primrose 1975; McAthey & Kilbey 1976; Wymer 1981).

The main working parameters are the dilution rate (D) and the cell generation time (t_d) which can be calculated as follows: $D = F/v$, where F is the flow rate of the fresh medium measured in ml/h and v is the working volume of culture in the growth chamber. $t_d = 0.693/\mu$. For routine purposes we use a t_d of 6.9 h.

Growth chamber

To reduce the possibility of wall growth, a well-designed growth chamber minimizes the contact area between the culture and the vessel which contains it. Similarly, the number of tubes, stirrers, etc., submerged in the culture should be kept to a minimum. At low cell concentrations the incoming gas supply usually provides sufficient turbulence to maintain the cells in suspension. At higher cell concentrations or in growth chambers with working volumes in excess of 500 ml, additional means of agitation are required. While commercial chemostats usually incorporate a paddle system in the growth chamber for this purpose, a magnetic stirrer will serve equally well.

Nutrient supply system

Ideally the medium reservoir should be large enough to contain sufficient growth medium for the entire growth period. In practice, however, it is usually necessary to refill this reservoir from time to time during the course of an experiment. A

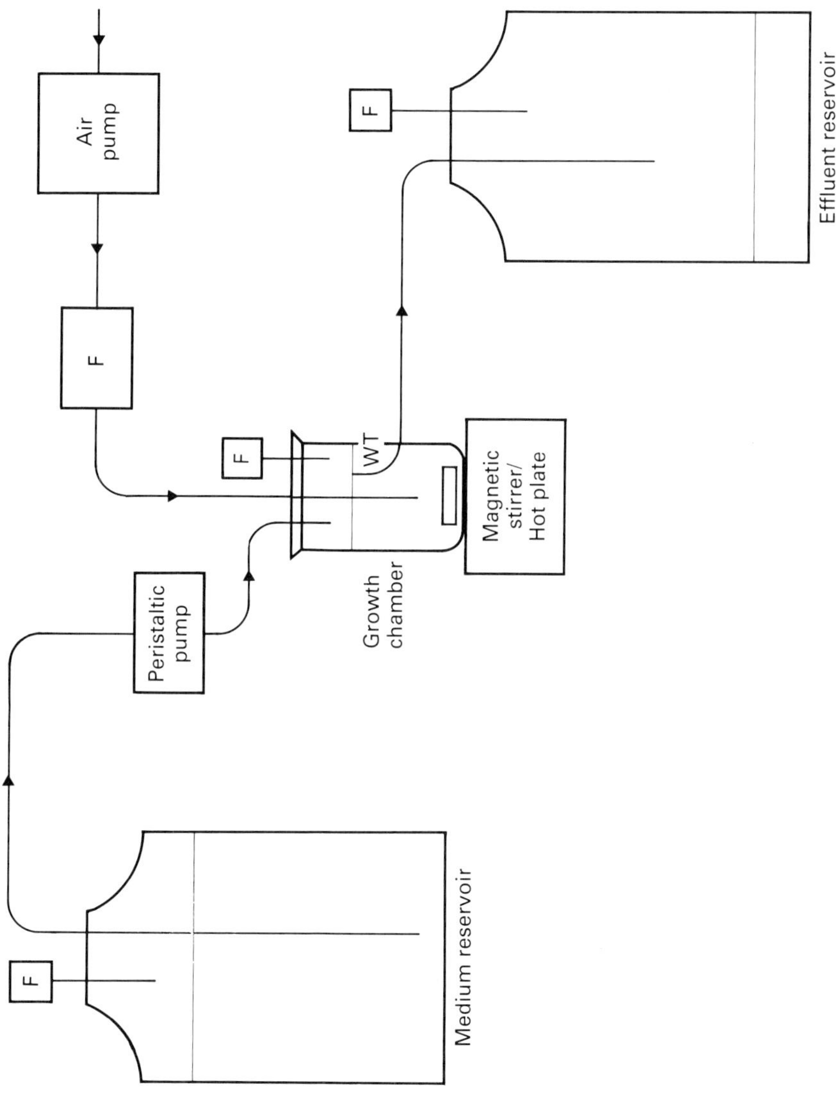

FIG. 2. Diagrammatic representation of a chemostat. F, filter; WT, weir tube.

controlled flow of growth medium from the medium reservoir to the growth chamber can be achieved either by the use of a Mariotte bottle (for details see Kubitschek 1970; Primrose 1975) or, more usually, by the use of a peristaltic pump. This is potentially the most expensive component of a chemostat; however, we have found the variable speed multi-mini pump marketed by Schuco Scientific Ltd to be adequate for this purpose.

Gas supply system

This is required for the growth of both aerobic and anaerobic organisms, as a positive pressure within the growth chamber reduces the chance of contamination. For the growth of *Sch. pombe*, filtered air is pumped into the chamber by means of a Hyflo air pump (Gallenkamp Ltd). The air outlet tube in the growth chamber should be plugged with cotton wool to prevent aerial contamination, as should the air inlet tube in the medium reservoir and the air outlet tube in the effluent reservoir.

Overflow system

We have found the most successful overflow system to comprise a weir tube leading from the surface of the culture, through the side wall of the growth chamber to an effluent reservoir.

All parts of the chemostat are connected with silicone tubing as this is more durable than natural rubber and does not release toxic substances on autoclaving.

Cultivation of *Sch. pombe*

Defined growth medium

The medium we use for the growth of *Sch. pombe* in continuous culture is Edinburgh minimal medium number 2 (EMM2) (modified from Mitchison 1970) which contains (per litre): glucose, 5 g; NH_4Cl, 5 g; NaH_2PO_4, 300 mg; CH_3COONa, 1 g; KCl, 1 g; $MgCl_2$, 500 mg; Na_2SO_4, 10 mg; $CaCl_2$, 10 mg; inositol, 10 mg; nicotinic acid, 10 mg; calcium pantothenate, 1 mg; biotin, 10 μg; H_3BO_3, 500 μg; $MnSO_4.H_2O$, 400 μg; $ZnSO_4.7H_2O$, 400 μg; $FeCl_3.6H_2O$, 200 μg; $H_2MoO_4.H_2O$, 160 μg; KI, 100 μg; $CuSO_4.5H_2O$, 40 μg; citric acid, 1 mg.

It is convenient to keep the components, especially the salts, vitamins and trace elements, as concentrated stock solutions (salts at 50 × working strength and the vitamins and trace elements at 100 × working strength) and store these at 4°C until required.

Sterilization

Ideally the whole apparatus containing sufficient medium for the entire growth period should be sterilized as a unit in an autoclave. In practice, however, we find it more convenient to sterilize the component parts separately and then reassemble them before use.

Inoculation

At the beginning of an experiment the growth chamber is inoculated with a 48 h batch culture and, after overnight incubation to permit cell density to increase sufficiently, the medium supply system is connected and the peristaltic pump set at the required speed. Steady state, as judged by cell density, is generally achieved within 24–48 h.

Control of temperature, pH and oxygen concentration

As growth rates in chemostats are relatively independent of external conditions, in practice we do not find it necessary to monitor the pH or oxygen concentrations when growing *Sch. pombe* in chemostat culture. Care is taken, however, to maintain the temperature of the growth chamber at 32°C.

Isolation of Mutants

Sampling

For most purposes we remove twice daily samples from the overflow system. Culture samples are harvested by centrifugation and washed twice in Giese's salts vitamins buffer (modified as described by Clarke 1963) which contains (per litre): $(NH_4)_2SO_4$, 3 g; $MgSO_4.6H_2O$, 700 mg; NaCl, 500 mg; Ca $(NO_3)_2$, 400 mg; KH_2PO_4, 6.8 g; inositol, 10 mg; nicotinic acid, 10 mg; calcium pantothenate, 1 mg; biotin, 10 μg. Samples are then resuspended in buffer to a concentration of *c.* 1×10^8 cells per ml and 0.1 ml amounts plated on appropriate selective medium. Yeast extract agar (YEA) (Leupold 1955) is used as complete medium. This contains (per litre): yeast extract, 5 g; glucose (or glycerol for the isolation of chloramphenicol and erythromycin resistant mutants), 30 g; agar, 20 g. YEA is supplemented as required with 2 μg/ml trichodermin (Leo Pharmaceuticals Ltd, Ballerup, Denmark), 100 μg/ml cycloheximide, 1 mg/ml chloramphenicol or 1.5 mg/ml erythromycin (all obtained from Sigma, London). These antibiotics are filter sterilized prior to their addition to autoclaved medium. In order to estimate viable counts, each cell suspension is also suitably diluted and 0.1 ml amounts plated on unsupplemented YEA.

Both selective and viable count plates are incubated at 32°C. Colony counts are generally made after 72–96 h, except for YEA plates supplemented with chloramphenicol when colony counts are made after 10–12 days.

Spontaneous mutation

In order to determine the background level of spontaneous mutation to resistance to the various antibiotics or metabolite analogues under test, continuous cultures of *Sch. pombe* should be grown in unsupplemented EMM2 and samples processed as described above.

Induced mutation

In order to study the effect of either known or potential chemical mutagens, continuous cultures of *Sch. pombe* are grown in EMM2 supplemented with the appropriate agent. As it is obviously important to establish that the chemical under test does not reduce cell viability, preliminary batch culture experiments have to be carried out to determine its maximum non-lethal concentration. Mutants are then

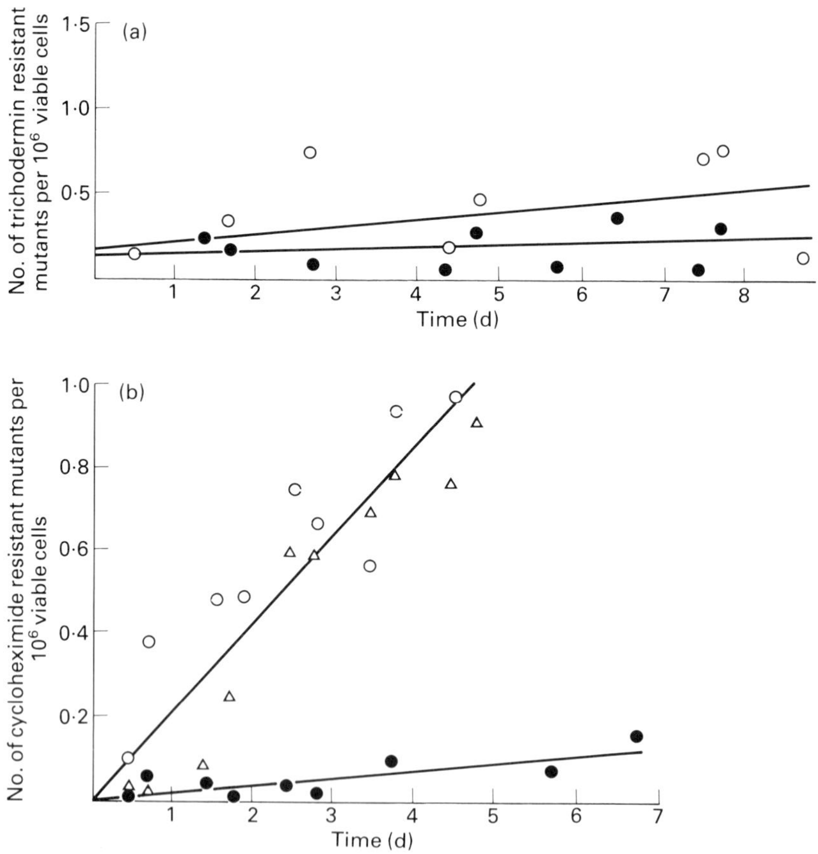

FIG. 3 (a). Accumulation of spontaneous (●) and EMS-induced (○) trichodermin-resistant mutants. (b). Accumulation of spontaneous (●) and EMS-induced (○, first experiment; △, second experiment) cycloheximide-resistant mutants.

isolated, as described above, from chemostats containing concentrations below this limit.

Data analysis

We find it convenient to express mutation frequencies as number of mutants per 10^6 viable cells. In an individual chemostat experiment the increase in mutant numbers can be followed by plotting this value against the time at which a particular sample was taken (see Fig. 3). Least squares linear regression lines are then calculated and fitted in the standard manner.

Comparison of the rate of accumulation of mutants in the presence and in the absence of the test chemical permits a quantification of its mutagenic activity. In practice we find that such a comparison can be made with reasonable accuracy after a minimum of 7 days sampling at steady state.

Examples of Data Obtained by Application of Technique

Ethyl methanesulphonate-induced mutation

The alkylating agent ethyl methanesulphonate (EMS) has been shown to induce mutations in many systems including continuous cultures of *E. coli* (Nestmann 1975). It is used by some workers as a positive control in the Ames test (for review see Mohn 1981). In our screening system, at a concentration of 120 μg/ml (the highest non-lethal concentration) EMS did not significantly increase the frequency of trichodermin resistant mutations above the spontaneous level (Fig. 3a, Table 1). At this concentration, however, EMS induced mutations to resistance to cycloheximide (Fig. 3b, Table 1), to chloramphenicol and to erythromycin (Table 1) at several times the respective spontaneous mutation rates.

Metal-induced mutation

The potential hazard to human health caused by heavy metals in the environment has been the subject of much discussion. The genotoxicity of some metals is now well established, but in other cases available evidence is less conclusive. For example, current data on the mutagenicity of lead and manganese in short-term bacterial tests are somewhat contradictory, and, although certain nickel compounds are among the best-documented experimental carcinogens, nickel has usually been found to be negative in these tests (Heck & Costa 1982; Hansen & Stern 1984). In our screening system, however, in which cytotoxicity, one of the main complications of the bacterial tests, is minimized, we have been able to demonstrate that, when present individually in the growth medium, lead, nickel and manganese induce mutations to trichodermin resistance and to chloramphenicol resistance, although only manganese significantly increases the frequency of cycloheximide-resistant mutations above the spontaneous level (Table 1). The particular sensitivity of yeast

TABLE 1. *Ratios of induced antibiotic resistant mutation rates to the corresponding spontaneous mutation rates*

		Ratio for:			
Mutagen(s)	Concentration (μg/ml)	Trichodermin resistance	Cycloheximide resistance	Chloramphenicol resistance	Erythromycin resistance
EMS	120	1.8*	13.1	22.3	20.0
Lead nitrate	40	7.3	1.5*	9.9	NT
Nickel nitrate	10	3.2	0.7*	7.9	NT
Manganese chloride	40	32.9	2.4	15.9	NT
Lead nitrate plus Nickel nitrate	40 + 10	17.8	2.6	18.1	NT
Lead nitrate plus Manganese chloride	40 + 40	63.6	4.1	27.5	NT

NT = Not tested.
*Not significant, induced mutation yield less than double the spontaneous level.

systems to weak mutagens and promutagens without the requirement for exogenous metabolic activation has also been noted by others (Nestmann *et al.* 1982; Nestmann & Lee 1983).

In light of suggestions that lead may act as a co-mutagen in humans accidently exposed to mixtures of heavy metals (Gebhart 1984), we decided to monitor the kinetics of mutation accumulation in *Sch. pombe* simultaneously challenged with lead and nickel or lead and manganese. When EMM2 was doubly supplemented with lead and nickel, additive increases in induced mutation yields were observed when cycloheximide resistance and chloramphenicol resistance were monitored, while a synergistic response was observed with trichodermin resistance (Table 1). Essentially similar kinetics of mutation accumulation were also observed when EMM2 was doubly supplemented with lead and manganese (Table 1). The genotoxic synergism observed in these studies is a particular cause for concern. A similar synergism has been reported between cadmium (another heavy metal which gives equivocal results in bacterial assays of mutagenic activity) and two nitrosamines (Mandel & Ryser 1984). This therefore suggests the possibility that weak mutagens, such as heavy metals, might become important secondary sources of genetic damage when acting synergistically with other agents.

Conclusions

We have developed a screening test for environmental mutagens employing the fission yeast, *Sch. pombe*. This system monitors the increase in mutation frequencies at selected loci in chemostat-grown cultures and can be used to detect chronic levels of mutagenic activity which might go undetected in short-term bacterial tests such as the Ames test.

Our experimental data show the degree of mutation induction varies with both the inducing agent(s) used and the individual antibiotic resistance scored. This emphasizes the danger of relying, as in many current tests designed to detect mutagenic activity, on a single forward or reverse mutation step.

Acknowledgments

We are grateful to Professor C. Auerbach, Dr B. J. Kilbey and Dr J. F. T. Spencer with whom we have had many stimulating discussions. We are also grateful to Mrs D. M. Spencer for the loan of a multiple fermenter, to Dr W. O. Godtfredsen for the gift of the trichodermin and to Dr R. Matthews for his assistance with the statistical analysis. This work was financed by the Polytechnic of North London.

References

Ames, B. N. 1979 Identifying environmental chemicals causing mutations and cancer. *Science, N.Y.* **204**, 587–593.

Bauer, K. H. 1928 *Mutationstheorie der Geschwulstentstehung.* Berlin: Springer-Verlag.

Cairns, J. 1978 *Cancer: Science and Society.* San Francisco: W. H. Freeman & Co.

Clarke, C. H. 1963 Suppression by methionine of reversions to adenine independence by *Schizosaccharomyces pombe. Journal of General Microbiology* **31**, 353–363.

Egel, R., Kohli, J., Thuriaux, P. & Wolf, K. 1980 Genetics of the fission yeast *Schizosaccharomyces pombe. Annual Reviews of Genetics* **14**, 77–108.

Egilsson, V., Evans, I. H. & Wilkie, D. 1979 Toxic and mutagenic effects of carcinogens on the mitochondria of *Saccharomyces cerevisiae. Molecular and General Genetics* **174**, 39–46.

Gale, E. F., Cundliffe, E., Reynolds, P. E., Richmond, M. H. & Waring, M. J. 1981 *The Molecular Basis of Antibiotic Action.* London: John Wiley & Sons.

Gebhart, E. 1984 Chromosome damage in individuals exposed to heavy metals. *Toxicological and Environmental Chemistry* **8**, 253–266.

Gutz, H., Heslot, H., Leupold, U. & Loprieno, N. 1974 *Schizosaccharomyces pombe. Handbook of Genetics*, Vol. 1, ed. King, R. C., pp. 395–446. New York: Plenum Press.

Gygax, A. & Thuriaux, P. 1984 A revised map of the fission yeast *Schizosaccharomyces pombe. Current Genetics* **8**, 85–92.

Hansen, K. & Stern, R. M. 1984 A study of metal-induced mutagenicity *in vitro* and *in vivo. Journal of the American College of Toxicology* **3**, 381–430.

Heck, J. D. & Costa, M. 1982 *In vitro* assessment of the toxicity of metal compounds. II. Mutagenesis. *Biological Trace Elements Research* **4**, 319–330.

Kubitschek, H. E. 1970 *Introduction to Research with Continuous Cultures.* New Jersey: Prentice-Hall.

Leupold, U. 1955 Versuche zur genetischen Klassifizierung adeninabhängiger Mutaten von *Schizosaccharomyces pombe. Archiv Julius Klaus-Stiftung Vererbungsforschung, Zürich* **30**, 506–516.

Loprieno, N., Barale, R., von Halle, E. S. & von Borstel, R. C. 1983 Testing of chemicals for mutagenic activity with *Schizosaccharomyces pombe.* A report of the US Environmental Protection Agency Gen-Tox Program. *Mutation Research* **115**, 215–223.

Mandel, R. & Ryser, H. J.-P. 1984 Mutagenicity of cadmium in *Salmonella typhimurium* and its synergism with two nitrosamines. *Mutation Research* **138**, 9–16.

McAthey, P. & Kilbey, B. J. 1976 Trichodermin: a selective agent for screening forward mutations in continuous cultures of *Schizosaccharomyces pombe. Biologisches Zentralblatt* **95**, 415–421.

McAthey, P. & Kilbey, B. J. 1977 Mutation in continuous cultures of *Schizosaccharomyces pombe.* I. Dependence of the kinetics of mutation accumulation upon the growth-limiting nutrilite. *Mutation Research* **44**, 227–234.

McAthey, P. & Kilbey, B. J. 1978 Mutation in continuous cultures of *Schizosaccharomyces pombe.* II. Effect of amino acid starvation on mutational response and DNA concentration. *Mutation Research* **50**, 175–180.

McCann, J., Choi, E., Yamaski, E. & Ames, B. N. 1975 Detection of carcinogens as mutagens in the *Salmonella*/microsome test: Assay of 300 chemicals. *Proceedings of the National Academy of Sciences, U.S.A.* **72**, 5135–5139.

Mitchison, J. M. 1970 Physiological and cytological methods for *Schizosaccharomyces pombe. Methods in Cell Physiology*, Vol. 4, ed. Prescott, D. M., pp. 131–165. New York: Academic Press.

Mohn, G. R. 1981 Bacterial systems for carcinogenicity testing. *Mutation Research* **87**, 191–210.

Monod, J. 1950 La technique de culture continué. Théorie et applications. *Annales de l'Institut Pasteur* **79**, 390–410.

NESTMANN, E. R. 1975 Mutagenesis by nitrosoguanidine, ethyl methane-sulphonate and mutator gene *mut H* in continuous cultures of *Escherichia coli*. *Mutation Research* **28**, 323–330.

NESTMANN, E. R. & LEE, E. G.-H. 1983 Mutagenicity of constituents of pulp and paper mill effluent in growing cells of *Saccharomyces cerevisiae*. *Mutation Research* **119**, 273–280.

NESTMANN, E. R., STEPHEN, E. R., KOWBEL, D. J. & NASIM, A. 1982 Differential survival as an indicator of potential mutagenicity using repair deficient strains of *Saccharomyces cerevisiae* and *Schizosaccharomyces pombe*. *Canadian Journal of Genetics and Cytology* **24**, 771–775.

NOVICK, A. & SZILARD, L. 1950 Description of the chemostat. *Science, N.Y.* **112**, 715–716.

PHIPPS, J., NASIM, A. & MILLER, D. R. 1985 Recovery, repair and mutagenesis in *Schizosaccharomyces pombe*. *Advances in Genetics* **23**, 1–73.

PRIMROSE, S. B. 1975 Inexpensive chemostat constructed from standard glassware. *Laboratory Practice* January, 88–89.

WYMER, P. E. O. 1981 The use of chemostat in school biology. *Journal of Biological Education* **15**, 259–262.

The Testing of Non-metallic Materials for Use in Contact with Potable Water, and the Inter-relationships with In service Use

J. ASHWORTH AND JENNIFER S. COLBOURNE*
Albury Laboratories Limited, The Old Mill, Albury, Guildford, Surrey GU5 9AZ, and
**Regional Laboratory Services, Thames Water Authority, 177 Rosebery Avenue, London EC1R 4TP*

Introduction

Traditionally the range of materials which come into contact with potable water was limited to concrete, metals and organic substances derived generally from natural sources. Since the mid-1960s materials derived from the petrochemical industry have been recommended and used on a wide scale in water systems.

Many of these materials, both traditional and modern, can exert deleterious effects (Schoenen & Scholer 1985; Colbourne 1985) such as unacceptable tastes, turbidity or leaching of constituents (either metals or organic substances) on water quality. Furthermore, many products stimulate microbial growth to such an extent that slimes appear on the surface of the material and the numbers of organisms in the water increase to levels that exceed recommended hygienic standards (Anon. 1982a, 1983).

The need to prevent the occurrence of such detrimental effects by assessing the suitability of materials before usage was formally recognized in 1975 when the National Water Council* introduced a test protocol (Fig. 1) based on methods developed by Thames Water (Burman & Colbourne 1977; Colbourne & Brown 1979). A decade of experience with the use of this protocol has resulted in a number of modifications designed to reflect the performance of the material under the most exacting service conditions and to improve the predictive value of the results. The techniques are described in full in a British Standard Draft for Development (Anon

* The National Water Council was abolished by Parliament in 1983 and the scheme is now administered on behalf of the U.K. water industry by the Water Research Centre.

Industrial Microbiological Testing
0-632-01793-7

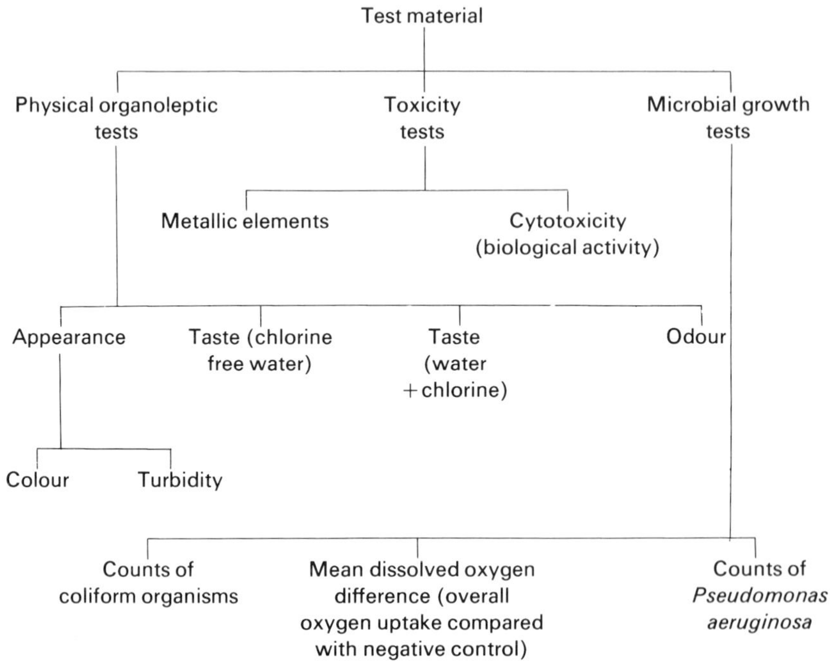

FIG. 1. Test protocol. Note: (1) Metallic materials fittings for metallic elements only. (2) Tests performed at 21 ± 3°C for cold water use and 85 ± 2°C for hot water use (microbial growth test excepted).

1982b) whilst the assessment criteria and more recent modifications are given in a Water Research Centre Information and Guidance Note (Anon 1984).

Implicit in any investigation into the causes of deterioration of water quality in building services, treatment plants, distribution networks or water fittings is the need for a detailed knowledge of the materials used and their performance in tests of this type. Of the tests shown in Fig. 1, the microbial growth and taste tests have probably the greatest significance in resolving practical problems that occur in customers' premises.

Microbiological Test Method

Although numerous techniques have been used to investigate field problems, in general these fall into one of four broad categories: (1) exposure of the material to agar plates seeded with pure cultures of micro-organisms; (2) immersion of the material in a single static volume of water containing micro-organisms derived from pure cultures; (3) exposure of the material to potable water under service conditions; (4) immersion of the material in potable water under dynamic conditions within the laboratory, usually with an initial inoculum of natural aquatic micro-organisms.

Application of the tests in each category has yielded valuable data but only those methods in category (4) have proved suitable for use as predictive tests (Schoenen & Scholer 1985).

The method developed by Burman & Colbourne (1977) and as modified subsequently by Colbourne & Brown (1979) forms the basis of the test protocol (Anon. 1982b). The method involves the complete immersion of the specimen in dechlorinated potable water containing initially a mixture of natural organisms derived from a surface water suitable for drinking water abstraction. This water is replaced with fresh uninoculated potable water twice weekly and the growth potential of the material assessed by direct counts and dissolved oxygen measurements in the water after each incubation period. Results are compared with a negative (glass) control. The test extends from 6 to 8 weeks.

Since 1975 three nationally accredited laboratories have conducted a series of inter-laboratory trials, in addition to evaluating over 3500 materials in common use, and the suitability of these materials has been ranked against the published criteria (Anon. 1984). The main findings can be summarized as follows:

Sample preparation

Samples of materials intended for application 'on-site' rather than under factory conditions may give variable results unless the test specimens are prepared and cured under carefully controlled environmental conditions (e.g. temperature, humidity, time). Control of these conditions has been shown to be critical in service situations also (Parkinson *et al.* 1983; Schoenen & Scholer 1985). For similar reasons samples specially prepared for test purposes may give different results from production materials.

Surface area to volume ratio

The test protocol allows a wide tolerance ($\pm 10\%$) on the ratio of 15,600 mm^2 sample surface area to 1000 ml soak water, since many specimens submitted for test can be very irregular in shape. A tolerance of $+9\%$ to -6% does not appear to exert any obvious influence on the number of variable results within or between laboratories. Wherever practical, however, the precise ratio should be maintained.

Test water

In order to measure the ability of a material to support microbiological growth under test conditions, all rate-limiting factors, other than those associated with the test specimen, are standardized. One important factor is the inorganic nutrient content of the water. The influence of the concentrations of orthophosphate and nitrate nitrogen in the test water has been studied, and minimum levels specified (Table 1). Potassium concentration has been studied also but found to be of little relevance.

Test inoculum

Schoenen & Scholer (1985) have identified the primary limitation of an inoculum based on laboratory cultures as a lack of correlation between test results and in-service performance. Questions have been raised also regarding mixed natural inocula. An extensive Water Research Centre trial using eight different sources of natural inocula and test waters found no obvious differences in results when using different inocula or waters in various laboratories (Warren 1985). More recently, two laboratories assessed the same material in duplicate using two different inocula sources (Table 2). These results suggest that comparable test results can be obtained using natural inocula, which have the advantage of giving a more reliable indication of in-service performance (Schoenen & Scholer 1985).

Test criteria

Colbourne & Brown (1979), using fifty different materials, demonstrated a clear correlation between the microbial growth response as measured by dissolved oxygen uptake and actual microbiological service problems as defined by Schoenen & Scholer (1985) (Table 3).

Data obtained from a more recent trial funded by the Department of Trade and Industry and based on two chemically consistent reference materials, glass and paraffin wax (representative of either end of the microbial response range) provide a quantitative measure of both inter- and intra-laboratory variability (Table 4).

Inspection of these data indicates that the intra-laboratory variability (average ± 0.6 mg/l) is similar to the inter-laboratory variability (average ± 0.4 mg/l). This suggests that a tolerance of ± 0.5 mg/l should be applied to the assessment criterion in order to take account of the inherent variability of the protocol and individual laboratory performance. In order to relate the criterion to service performance, a value of 2.3 ± 0.6 mg/l is recommended.

More recently a comparative study of the performance of this method with that developed by Schoenen & Scholer (1985) has been undertaken in the laboratories of both Thames Water Authority, London, and the Institute of Hygiene, University of Bonn. Duplicate samples of eleven polyvinyl chloride materials were tested by both organizations and 100% agreement was obtained with regard to the final assessment of the suitability of these materials for use in contact with potable water (Schoenen & Colbourne 1987).

TABLE 1. *Uptake of inorganic nutrients from test water using glass and paraffin wax control materials (results from an inter-laboratory trial)*

	Laboratory A			*Laboratory B*			*Laboratory C*		
	PO_4 (mg/l)	NO_3–N (mg/1)	K (mg/l)	PO_4 (mg/l)	NO_3-N (mg/l)	K (mg/l)	PO_4 (mg/l)	NO_3-N (mg/l)	K (mg/l)
1. Initial inorganic nutrient levels in the test waters									
Normal test water	1.9	6.5	4.8	0.01	2.7	2.0	0.01	0.1	0.2
Supplemented test water	> 2.0	> 5.0	> 5.0	> 2.0	> 5.0	> 5.0	> 2.0	> 5.0	> 5.0
2. Inorganic nutrients removed during incubation with test materials									
Normal test water									
With glass negative control material	0.1	0.4	0.0	0.0	0.0	0.0	0.0	0.02	0.0
With paraffin wax positive control material	0.8	4.4	0.0	0.0	0.2	0.0	0.0	0.09	0.0
Supplemented test water									
With glass negative control material	0.1	0.4	0.0	0.4	1.5	0.0	0.0	0.0	0.0
With paraffin wax positive control material	0.8	4.4	0.0	1.3	3.5	0.0	0.2	2.6	0.0

TABLE 2. *Assessment by two laboratories of the microbial growth response of duplicate samples of an elastomeric material using two different natural inocula*

	Mean dissolved oxygen difference* (mg/l)	
	Laboratory A	Laboratory B
River inoculum A (River Thames)	5.0	5.2
River inoculum B (River Wey)	4.8	4.9

* Colbourne & Brown (1979).

TABLE 3. *Relationship between microbial growth and mean dissolved oxygen values of fifty different materials*

Laboratory system (Colbourne & Brown 1979)	In-service problems (Schoenen & Scholer 1985)	
	Degree of microbial growth observed under 'in-service' conditions	
Range of mean dissolved oxygen difference* (mg/l) exhibited in defined laboratory tests	Surface growth	Increased bacterial counts in water
0.0–2.1	−	−
2.2–3.4	+	−
3.5–8.0	+	+

* Colbourne & Brown 1979

TABLE 4. *Inter- and intra-laboratory variability in the assessment of the microbial growth associated with paraffin wax and glass reference materials*

Reference material	Mean values (see key) for duplicate tests in each laboratory (mg/l)			Overall mean of six tests in three laboratories (mg/l)
	Lab A	Lab B	Lab C	
Paraffin wax (melting point 52–54°C*)	6.7 ± 0.4	7.3 ± 0.3	6.2 ± 0.1	6.7 ± 0.6
Glass†	7.9 ± 0.5	8.9 ± 0.6	7.9 ± 0.2	8.2 ± 0.5

Key: (1) Positive control (*) = mean dissolved oxygen differences.‡ (2) Negative control (†) = mean dissolved oxygen values.

‡ Colbourne & Brown (1979).

Materials Performance

Colbourne (1985) and Schoenen & Scholer (1985) have shown that microbial growth, either determined in the laboratory or observed under service conditions, is a material specific phenomenon. Two basic mechanisms appear to exist: (1) Organic constituents, which act as nutrients, leach from the material into the water over a limited period of time varying from a few days to several years. Microbial growth occurs in the water. (2) Organic compounds at the material surface provide a source of nutrients for certain of the micro-organisms that colonize the surface. A mixed population of micro-organisms develops at the material/water interface; once established this biofilm will persist indefinitely.

In both cases the phenomenon can be related to the chemical composition of the material and in general the presence of particular additives or processing contaminants. Exceptionally the polymer itself is subject to biodegradation. A comparison of material performance according to polymer type highlights these findings (Table 5). A more detailed survey of material performance is given by Colbourne (1985).

Replicate testing of the same material, and exchange of identical specimens

TABLE 5. Overall performance of material types in the microbial growth tests involving measurement of oxygen uptake attributable to microbial growth

	No. of samples.	Percentage of samples giving an MDOD* greater than 2.3 mg/l
Elastomeric materials		
Natural rubber	14	100
Synthetic rubbers	606	67
Silicone rubbers	19	0
Silicone sealants	11	25
Polyester elastomers	8	0
Polyurethane elastomers	17	69
Polysulphide elastomers	9	100
Plastic materials		
Non-reinforced plastics	387	2
Reinforced plastics	61	12
Plasticized materials	47	68
Coating formulations	120	28
Miscellaneous products including jointings, adhesives, lubricants, cementitious products	107	38

* Mean dissolved oxygen difference (Colbourne & Brown 1979).

between laboratories, has yielded results that suggest that where variable responses are experienced these are associated with the uneven distribution of additives throughout the matrices of certain polymer types.

Although further work is continuing, there is considerable evidence to support the continued use of the current protocol for the purpose of ranking materials in order of their suitability for use in contact with potable water. The current list of suitable materials for use in contact with potable water is published biannually by the Water Research Centre (Anon. 1986a).

Investigation of Problems

Water quality problems are common and when they arise the consumer's typical response can be: (1) ignore the problem and do nothing—hope it will go away; (2) panic—when it doesn't; (3) admit there is a problem and gather information; (4) call in outside help (Allsop 1982).

Problems usually turn out to be attributable to: (1) a single cause; (2) multiple causes; (3) 'red herrings'—problem has a non-biological cause; (4) inherently insoluble factors; (5) associated with problems, long since resolved (Seal 1982).

Complaints about the quality of drinking water are summarized in Table 6. Complaints about unacceptable tastes are the most common with about 70% of these being of the mouldy, musty and earthy tastes caused either by microbial growth or an unsuitable material. The remainder are associated with ill-health, the mineral composition or disinfection of water, and the condition of pipework.

TABLE 6. *Percentage breakdown of a typical annual list of consumer complaints about water quality* (Greater London area)*

	Percentage
Taste	58
Dirty water	16
Aeration	12
Illness	6
Microbiological quality	2
Animalcules	2
Oil/grease/scum/hardness	2
Others	2

* Data derived from Thames Water Authority consumer complaint monitoring system.

Investigation Procedure

Irrespective of the type of problem an investigation should include the following steps:

Collection of relevant information

(1) Identify the source of the water supply to the premises. For public supplies contact the water undertaking for information on average analysis and knowledge of local water quality. For private supplies obtain analytical data and details of treatment from the owner. If none available then contact the Local Authority Environmental Health Officer who is responsible for monitoring the quality of private supplies.

(2) Obtain a simple schematic flow diagram of the water services inside the building. The following features, if present, must be highlighted: (1) location of incoming mains supply and/or private source; (2) booster vessels, pumps and break tanks; (3) cisterns; (4) materials of construction of pipework; (5) types of fittings, e.g. taps, showers, valves; (6) calorifiers, water heaters, etc.; (7) cooling towers, heating circuits, industrial services; (8) air conditioning systems, humidifiers.

(3) Perform a site survey. Starting at the point of the incoming supply, follow the route of the services relevant to the problem under investigation and note on the diagram the condition of the pipes, the jointing method(s) used, the presence of lagging, sources of heat and degree or adequacy of protection of cisterns. Look for disconnected fittings, 'dead-end' and cross-connections with other services.

Identify suitable sampling points at the following locations: incoming supply, cistern and/or calorifier, an outlet at the distal point. If the problem is associated with a particular fitting, sample the water entering and leaving the fitting. It is of particular importance to ensure that the sampling point represents the water under investigation. Draw off points at water meters will not, for example, give a representative sample of the incoming mains supply since the drain point is on a spur and will contain stagnant water.

Collection of samples

Collect samples in containers prepared in accordance with the proposed method of analysis (Anon. 1983; Standing Committee of Analysts). Collect samples in the following sequence from each point ensuring first that the outlet has not been flushed. The external surface and rim of the outlet should be clean and free from deposits. (1) bacteriology (minimum 400 ml, up to 1000 ml); (2) metals (500 ml); (3) residual chlorine (10 ml, on-site measurement); (4) taste (100 mls), on site assessment; (5) temperature—place a thermometer in the water flow and allow the outlet to discharge until the temperature stabilizes; record the initial and final temperatures and the time taken to reach the latter; (6) chemical (1000 ml, appearance and non-metals chemistry); (7) metals (500 ml); (8) bacteriology (minimum 400 ml, up to 10 litres); (9) taste—on-site assessment and collect 100 mls for return to the laboratory.

The initial samples (nos. 1–4) are designed to indicate contamination within the sample point/fitting and the final samples (nos. 6–9) represent the quality of the water being supplied to the fitting. Collect in sterile containers actual specimens, deposits, etc., as appropriate. Record the location, date and time of each sample.

Analytical procedures

(1) Microbiology. Routinely tests should include those for coliform bacteria (Anon. 1983); microfungi (Burman & Colbourne 1977); *Pseudomonas aeruginosa* (Anon. 1983); plate counts—22°C/72 h, 37°C/48 h (Anon. 1983). Where relevant, tests for other organisms (enteroviruses, algae, animalcules (Anon. 1982a), salmonellae, streptococci, other pseudomonads, aeromonads and *Legionella pneumophila*) should be included.

(2) Direct microscopic or chemical examination of specimens.

(3) Physical/chemical properties: colour, turbidity, electrical conductivity, appearance. Chemical analyses relevant to water source and treatment, e.g. for chlorides, total hardness, alkalinity, nitrate, nitrite, sulphate, total dissolved solids, phosphate.

(4) Determination of metals content: lead, zinc, copper, iron.

(5) Determination of other substances, where relevant: these may include polyaromatic hydrocarbons, phenols, other metals, sodium, fluoride, trace elements.

Survey of fittings

Determine whether these meet the requirements of the Model Water Byelaws (Anon. 1986a) by reference to the guide (White & Mays 1986) and the Directory (Anon. 1986b). For suspect components of fittings, collect specimens from site and obtain, if possible, fresh unused samples of materials/fittings. Test these in accordance with the protocol described previously (Fig. 1).

Recommendations for Remedial Measures

Recommendations will vary according to the type of problem. Examples of such recommendations include: (1) the replacement of the material/fitting with suitable alternative(s); (2) relocation of fitting or alteration of the supply pipework; (3) installation of proper sampling points; (4) insulation of the supply pipework; (5) protection of cisterns; (6) descaling of the pipework and/or fittings; (7) physical or chemical cleaning; (8) disinfection; (9) pasteurization of calorifiers; (10) installation of treatment device(s) or plant.

In many instances a combination of measures may be required to resolve problems. In general, remedial measures should be implemented one at a time and further sample analysis should be undertaken to verify that the action has proved successful before proceeding further. Advice on a long-term monitoring programme should be provided once the problem is overcome.

Discussion of Some Specific Matters Associated with Water Quality

The general principles dealt with above have found application in resolving water quality problems; in particular two areas currently generate much interest.

The Operation of Drinks Vending Machines (including Ice-Makers, etc.)

Many members of the public are suspicious about the products delivered by vending machines, often with good reason. New generations of electronics, including 'micro-processor' controls, appear to have revolutionized these machines, but with this increasing sophistication there has come a concomitant expectation from the public of higher quality drinks.

Drinks vending machines contain a complex water system and often a large number of different materials (plastics, rubbers and coatings). They are subject to intermittent water flow, with periods of stagnation lasting up to several days (e.g. weekends, bank holidays). Normally these machines do not incorporate an automatic mechanism for flushing water to waste after periods of non-use and often the machines are installed at the end of long runs of pipework.

The quality of the final vended product is related to the quality of the ingredients used, but even with high quality ingredients the drink may be unacceptable due to

TABLE 7. *Categories of complaint and their possible causes*

Complaint	Possible Causes
Off-taste or odour in drink	Contaminated or unsuitable ingredients Atmospheric contamination Unsuitable construction materials Inadequate cleansing or maintenance Unsuitable water supply to the machine
Particulate and foreign matter in drink	Contaminated ingredients Unsuitable construction materials or design Microbial growth
Illness	Pathogenic micro-organisms toxic metals, e.g. copper

incorrect dilution ratios, taints and odours picked up during storage or even as a result of microbiological growth within the ingredients.

Investigative procedure

Water quality. If the water entering a vending machine is of inferior quality then the final vended product will suffer. At the time of installation it is normal for little regard to be given to the quality of the water supply, and in large buildings it is often difficult, or well nigh impossible, to identify the source of the water or to follow the complete routing of the pipework.

No machine should ever be connected to a water supply drawn from a storage cistern unless it complies with the strict requirements for drinking-water cisterns (Anon. 1986b).

Irrespective of its source it is essential that the microbiological and chemical quality of the water complies with EC Directive 80/778/EEC as interpreted by the Department of the Environment (Anon. 1982a). A deterioration in microbiological quality of the water may occur if: (1) there are 'dead legs' (short lengths of spur pipes which supply a single fitting and are connected to the main feed distribution pipe); (2) the routing of the supply pipe is excessively long (lengths of up to 400 yards are not uncommon); (3) localized warming of the water in the pipework occurs.

In large buildings it is frequently the case that the water reaching the vending machine is at ambient temperature or higher. Recent examples investigated in office accommodation have exhibited water temperatures in the range 24–35°C. The uninsulated supply pipe had passed through either a plant room containing the central heating boiler, or ducting containing hot water and steam pipes. In an extreme example the water reaching the vending machine had a temperature of 35°C. Under these circumstances the water entering the machine can exhibit a total bacterial count in excess of 1×10^4 organisms/ml after incubation at either 22°C or 37°C. The EC Directive (Anon. 1982a) specifies that the temperature of the water at the point of consumption should not exceed 25°C.

Most machines do not provide a mechanism to flush water to waste and therefore water that has stood in the pipework overnight or during a weekend will be drawn into the machine at the time of the first morning vend. Investigative samples must be taken, therefore, from the water supply pipe without flushing the water to waste. Samples of vended cold water should be taken from the cup station. In 'hot drink only' machines water should be sampled from the supply pipe to the heater within the machine; in these machines, with adequate heating of the water (>70°C), no microbial problems should be associated with the dispensed hot drinks (Godfree *et al.* 1981). In a well-designed and correctly installed and maintained machine, the numbers of organisms in the water that has passed through the machine should not be more than 10 times that in the incoming supply. Where an increase of greater magnitude occurs then an investigation is required.

Siting of machine. Most machines contain storage (break) tanks that act as anti back-siphonage devices. These are open to the atmosphere, and may be subject to contamination if the machine is installed in a room where organic solvents are present in the atmosphere. These can dissolve in the water in the tank and support a copious growth of organisms (Poynter & Mead 1964). This latter problem is often encountered when machines are installed in automotive repair garages, printing works, paint factories, some health premises, hairdressers, etc. Additionally strong atmospheric odours can cause a direct chemical taint of the ingredients or water.

Ingredients. If these are suspect they should be examined microbiologically and compared with 'fresh' ingredients. Chemical analysis and organoleptic assessments may be necessary.

Materials of constructions. All food contact materials should comply with the requirements of Statutory Instrument 1978 No. 1927 (Anon. 1978). These requirements are less specific than those applied to materials in contact with potable water and this legislation does not specifically recommend a method for the assessment of microbial growth potentials.

If a material is suspected, a sequential set of water samples should be taken from within the machine, starting with one before any water treatment or filtration unit, one after this unit, and then one at various points up to the water heater, or in the case of cold vends, the cup station. Simple colony counts will identify the focal point of growth.

Mechanical design. Partly reconstituted or diluted ingredients can accumulate in crevices and on rough surfaces; visual examination of components from an in-service machine will readily identify this problem.

Many machines contain water treatment/filtration/softening units to 'polish' the incoming water (remove colour, certain tastes or suspended matter); softening is to protect heater components. These units can cause substantial increases in microbial numbers when used without regular maintenance, and in extreme cases a thousand-fold increase in the colony counts in the water may occur (Anon. 1980); to guard against this units should be clearly visible and, if installed within the machine, readily accessible.

Some case histories

Each problem arises usually from a unique set of circumstances, but generally includes the use of one or more unsuitable materials.

Case A. The translucent silicone rubber tubing delivering cold drinks to the cup station had a thick black film of fungi in its lumen. Flakes of this growth were found floating on the drinks, which had a 'musty' taste. The cause of the problem was a natural rubber component directly upstream of the tubing. The solution was to replace the natural rubber component with a component tested in accordance with the protocol described earlier and found to be satisfactory.

Case B. High bacteriological numbers in the water dispensed by a machine (total viable counts of greater than 10^5 per ml at both 22 and 37°C) were traced to the quality of the incoming water which was at 27°C, and to solvent vapours reaching the machine. Relocation of the machine cured the problem.

Case C. One design of machine caused different problems at various sites. In areas where the water supply had a low free residual chlorine content, the water had poor microbiological quality. In areas where there was a higher level of free residual chlorine drinks had a strong 'chlorophenolic' disinfectant type taste. All materials in contact with food and water were found to be satisfactory. However, a flexible PVC pipe, fitted to the overflow from the storage tank, emitted a soapy odour. In the laboratory it was suspended over but not in contact with the water containing an inoculum of micro-organisms and copious growth occurred in the water. Similarly when the pipe was suspended over chlorinated water, this developed a strong disinfectant taste. Thus both problems were caused by transfer of volatile components from the flexible pipe, via the atmosphere, to the water.

Water Systems Colonized with Legionella pneumophila

Legionella pneumophila, the most common cause of Legionnaires disease, was first recognized by McDade *et al.* (1977) and since then it has been established that legionellae are ubiquitous in natural waters and the environment as a whole. Investigations of outbreaks of the disease have focused attention on water systems inside large buildings, especially hot water services and cooling water circuits (Thornberry *et al.* 1984) which sometimes provide sites for prolific growth of legionellae. Where water containing high numbers of *L. pneumophila* is disseminated into the atmosphere in the form of an aerosol, susceptible individuals may be at risk of contracting either Legionnaires disease or Pontiac fever (Fitzgeorge *et al.* 1983).

The development of relatively straightforward techniques for the isolation of legionellae from waters (Fitzgeorge & Dennis 1983), coupled with a rising public concern about the safety of large buildings, has resulted in numerous *ad-hoc* samplings of building waters for the presence of these organisms. Unfortunately this non-systematic survey work has contributed little to our understanding of their ecology; however, four recent studies (Colbourne *et al.* 1984; Van der Kooij & Hijnen 1984; Voss *et al.* 1986; Schofield & Locci 1984) have demonstrated clearly that *L. pneumophila* is able to colonize pipes, fittings and structures within buildings in a manner akin to other common aquatic bacteria such as *Ps. aeruginosa* (Colbourne 1979).

As yet it is unclear whether *L. pneumophila* utilizes directly the organic constituents present in water or those derived from the plumbing materials. It is possible that the presence or absence of other species of aquatic organisms and the characteristics of the water may exert an influence on the extent to which any particular system becomes colonized. Circumstantial evidence from reported failures in methods of control, e.g. heat and disinfection (Thornberry *et al.* 1984) supports the theory that the organism multiplies within protected niches rather than in the circulating water itself.

Investigative procedure

When undertaking a survey for the presence of *L. pneumophila* in water systems it is of particular importance to establish in advance the purpose of the investigation. Usually this is to satisfy the building owners/occupiers that the water services are unlikely to pose any risk to health. The sampling protocol must be directed, therefore, towards establishing the focal point of contamination and features which provide conditions suitable for microbial growth so that the appropriate remedial method can be selected without need for further sampling or analysis. The protocol outlined above (Investigation of Water Quality Problems) is specially designed to distinguish between contamination of single fittings (e.g. showers) and contamination of water services (e.g. calorifiers, tanks). A full account of the importance of this type of sample protocol is given in both the study conducted at Kingston Hospital (Colbourne *et al.* 1984) and that of a cold water distribution network in Columbus (Voss *et al.* 1986). In cases where the presence of the organism was confined to fittings at the extremity of the water system, common methods of treatment with chlorine and heat have proved to be ineffective, costly, and in some instances, when carried out incorrectly, potentially hazardous to the occupants of the building.

In general, the simple bacteriological and chemical tests outlined in the protocol are sufficient to identify whether or not a water system is subject to gross microbiological contamination and whether or not any component parts or aspects of the installation are capable of acting as focal points for microbial growth. In the former case, the presence of legionellae is of much less importance than the potential hazard posed by the presence of enteric pathogens; particularly those derived from faecal matter and which are most likely to be detected by a test for faecal indicator organisms. Contamination of individual fittings that may contain sources of nutrients for growth in the form of accumulations of scale, rust and organic debris, or unsuitable materials such as washers, hoses, lubricants or jointing compounds, can be identified by comparison of simple bacteriological plate counts obtained before and after flushing of the fitting. These tests are inexpensive and it is possible to collect and analyse sufficient samples to give representative data on which to base a reliable judgment as to the quality of the water throughout the system.

Although the selection of sampling points should focus on those points capable of disseminating an aerosol of water into the atmosphere, samples from the bulk sources (e.g. calorifiers and tanks) should be collected for reference purposes. Once the appropriate remedial measures have been undertaken, further samples from a smaller number of selected points should be collected and tested (including for legionellae) if it will provide reassurance to the owner/occupants. Cooling waters, as opposed to drinking or domestic services, require regular rather than intermittent monitoring to ensure preservation of water quality, and in these instances a plot of bacteriological plate counts versus time over a period of several months should be obtained. As each system will be unique, either in its design or location, the schedule of water treatment (chemical dosing) and maintenance (draining down and physical cleaning) should be prescribed on the basis of such a quality profile. When setting out to investigate problems with these types of systems, past analytical results including chemical determinations, e.g. total dissolved solids, conductivity, pH, etc., should be examined first.

Outbreaks of Legionnaire's disease have captured the attention of the media and the public, thereby focusing attention on the importance of maintaining water quality within building services. This is much welcomed by water undertakings and Environmental Health Officers, who have long been aware of the problems that can result from a badly maintained or constructed water system. However, maintenance engineers, treatment specialists and investigating analysts alike should bear in mind that the indiscriminate application of hazardous treatment chemicals, and the failure to guard against faecal contamination or the use of materials capable of leaching toxic organic or metallic substances into the water, all pose a far greater potential risk to the health of building occupants and the general public than the presence of relatively low levels of *L. pneumophila.*

It is of vital importance that the results of survey findings are discussed with both the local water undertaking and Environmental Health Officer before commencing remedial treatment. This will ensure that discharges of chemically treated waters do not give rise to pollution of watercourses or impede the operation of local sewage treatment plants, and that the building owner is provided with up-to-date information on suitable materials of construction and chemicals, codes of practice, data on the quality of the water supply (whether public or private) and legal requirements.

Case histories

Although different water systems may become colonized with legionellae for a number of reasons, the following cases represent the most common situation encountered currently in domestic water services. They serve to highlight the importance of collecting representative samples in a manner that, in conjunction with appropriate analysis, will provide information that can be readily interpreted in terms of the design and operation of the building services. *Ad-hoc* sampling for *Legionella* alone will not provide such information.

Case A. Initially samples had been collected at random from showers without prior flushing, and *L. pneumophila* serotypes 1 and 6 were isolated from forty-two of ninety-four samples. No other analyses had been undertaken and the owners of the very complex commercial site were instructed to chlorinate all water services; this was done using a concentration of 5 mg/l free residual chlorine for 12 h. Further random samples proved to be positive when repeated on three separate occasions. At this stage a full water quality survey, as described previously, was undertaken and it rapidly became apparent that *L. pneumophila* could be isolated only from one of the four designs of shower (the only type not listed as meeting Water Byelaws requirements). *L. pneumophila* was not isolated from the cisterns, calorifiers or samples collected after thorough flushing of the showers. Simple bacteriological plate counts and testing of the shower components identified the thermostatic valve gasket as capable of supporting an abundant microbial slime. The gasket was replaced with a suitable alternative material, known to be satisfactory in the tests described previously, and *L. pneumophila* has not been isolated since.

The showers were all supplied with cold stored water and the temperature controls had been set to deliver water at a maximum of 50°C. No evidence of abnormal scale accumulation was observed in the services.

Case B. A Local Authority arranged for preflush samples of water to be collected for two showers and a drinking water tap connected to the mains within a swimming pool complex. All these samples contained *L. pneumophila* serotype 1. No other analyses were performed but the Local Authority notified the water undertaking because of concerns about the bacteriological quality of the public supply. A second sampling was arranged and on this occasion the sample protocol, as described earlier, was used to collect samples from the same drinking water tap and a stand pipe on the main in the street outside the premises. *L. pneumophila* serotype 1 subtype Olda (Watkins *et al.* 1985) was isolated from the 1 litre preflush tap sample, but legionellae were not isolated from the 10 litres after-flush tap sample or the 10 litre sample from the water main. Further typing of the three original isolates confirmed these to be *L. pneumophila* serotype 1 subtype Olda also.

The temperature of the water discharged from the drinking tap was 17°C higher than that of the supply main which was 10°C. The tap water contained 3225 μg/l of iron, 500 μg/l of copper and 1100 μg/l of zinc, whereas the mains water contained <10 μg/l of copper and zinc and only 151 μg/l of iron. The turbidity of the tap water was 0.5 NTU greater than that of the supply water. In all other respects the bacteriological and chemical quality of the samples was typical of the local public supply.

The complex had been subject to considerable renovation with new pipework laid approximately 3 years before the incident. The investigation cast considerable doubts about the manner in which the new services were commissioned; the design of the internal pipework did not conform with Water Byelaws in two respects—the use of mixed metal pipework (copper and galvanized steel) and inadequate protection against localized warming.

Case C. A Local Authority arranged for the collection of samples for *Legionella* analysis from a multioccupational residential premises tentatively associated with an unconfirmed case of Legionnaires disease. Preflush samples from the draincocks on the return hot water circuit were found to contain *L. pneumophila* serotype 3. A building water quality survey was undertaken, as described earlier, and paired samples were collected from the mains drinking water, domestic (stored) cold and hot water services in a manner designed to be representative of the circulating water. Additionally, samples were taken from the draincocks on the calorifier and return hot water feed circuit. *L. pneumophila* serotypes 2 and 3 were isolated only from preflush samples collected from draincocks situated on the hot water return feed. All samples collected from taps were free from legionellae in 10 litres.

The hot and cold water temperatures were generally $>50°C$ and $<20°C$ respectively at points of usage and the bacteriological and chemical quality of the water was satisfactory. However, the draincocks on the hot water return feed were located outside the buildings (the site comprised three blocks with the plant located centrally in one building) in a manner that could permit ingress of soil or surface water.

Conclusion

An investigation of a water system to identify the reasons for or source of microbiological contamination need not be a complex or lengthy procedure provided that care is taken to ensure that sufficient information is available before sample collection. Sampling protocols should be designed to identify the problem *and* to give sufficient information to recommend a practical solution as soon as the analytical results are available. Since every incident will present a unique environment, thorough site surveys, sample collection and analysis are important.

Acknowledgments

The authors wish to thank Mr A. Tebby, Managing Director, Albury Laboratories Ltd., and Mr. W. R. Harper, Managing Director, Thames Water Authority, for permission to publish this paper. We also thank the Department of Trade, Metrology and Standards Requirements Board who have granted permission to include in this report results from a recent interlaboratory trial funded by them.

In addition they would like to thank Dr P. J. Dennis and colleagues, PHLS Centre for Applied Microbiology and Research, Porton Down, Salisbury, Wiltshire, for typing the legionellae strains encountered in the case histories described. This work was undertaken as part of a collaborative PHLS CAMR/Thames Water Authority research project.

References

ALLSOP, D. 1982 Biodeterioration in practice: Part 1. Trouble shooting for industry. *Laboratory Practice* **31**, 220–222.

ANON. 1978 The Materials and Articles in Contact with Food Regulations 1978. *Statutory Instruments 1978 No. 1927*. London: Her Majesty's Stationery Office, and subsequent amendments.

ANON. 1980 Assessing the Effectiveness of Small Filtration Systems for Point-of-use Disinfection of Drinking Water Supplies. Health and Welfare Canada Publication 80-EHD-54, Ottawa.

ANON. 1982a EC Directive Relating to the Quality of Water Intended for Human Consumption (80/778/EEC). Department of the Environment Circular 20/82. London: Her Majesty's Stationery Office.

ANON. 1982b Specification of requirements for suitability of materials for use in contact with water for human consumption with regard to their effect on the quality of the water. Draft for Development 82:1982. London: British Standards Institution.

ANON. 1983 The Bacteriological Examination of Drinking Water Supplies 1982. *Reports on Public Health and Medical Subjects No. 71.* London: Her Majesty's Stationery Office.

ANON. 1984 Requirements for the Testing of Non-metallic Materials for use in Contact with Potable Water. *United Kingdom Water Fittings Byelaws Scheme Information and Guidance Note No. 5-01-02.* Slough: WRc Water Byelaws Advisory Service.

ANON. 1986a Water Fittings and Materials Directory 1986. Issue No. 2/86-87. Marlow: Water Research Centre/Unwin Brothers Ltd.

ANON. 1986b Model Water Byelaws (1986 edition). Department of the Environment, Scottish Development Department, Welsh Office; London: Her Majesty's Stationery Office.

BURMAN, N. P. & COLBOURNE, J. S. 1977 Techniques for the assessment of growth of microorganisms on plumbing materials used in contact with potable water. *Journal of Applied Bacteriology* **43**, 137–144.

COLBOURNE, J. S. 1979 The presence and significance of *Pseudomonas aeruginosa* in London's water supply. Ph.D. thesis, University of Surrey.

COLBOURNE, J. S. 1985 Materials usage and their effects on the microbiological quality of water supplies. In: *Microbial Aspects of Water Management. Journal of Applied Bacteriology Symposium Supplement* **59**, 47S–59S.

COLBOURNE, J. S. & BROWN, D. A. 1979 Dissolved oxygen utilisation as an indicator of total microbial activity on non-metallic materials used in contact with potable water. *Journal of Applied Bacteriology* **47**, 223–231.

COLBOURNE, J. S., PRATT, D. J., SMITH, M. G., HARPER, D. & FISHER-HOCH, S. P. 1984 Water fittings as sources of *Legionella pneumophila* in a hospital plumbing system. *Lancet* **i**, 210–213.

FITZGEORGE, R. B., BASKERVILLE, A., BROSTER, M., HAMBLETON, P. & DENNIS, P. J. 1983 Aerosol infection of animals with strains of *Legionella pneumophila* of different virulence: comparison with intraperitoneal and intranasal routes of infection. *Journal of Hygiene* **90**, 81–89.

FITZGEORGE, R. B. & DENNIS, P. J. 1983 Isolation of *Legionella pneumophila* from water supplies: comparison of methods based on the guinea-pig and culture media. *Journal of Hygiene* **91**, 179–187.

GODFREE, A. F., BOSLEY, M. & JONES, F. 1981 The use of drink vending machines with contaminated water supplies: the potential risks to public health. *Journal of the Institution of Water Engineers and Scientists* **35**, 80–87.

MCDADE, J. E., SHEPARD, C. C., FRASER, D. W., TSAI, T. R., REDUS, M. A. & DOWDLE, W. R. 1977 Legionnaires disease: isolation of a bacterium and demonstration of its role in respiratory disease. *New England Journal of Medicine*, **297**, 1197–1203.

PARKINSON, R. W., WARREN, I. C. & NORTH, K. J. 1983 Recent investigations of the epoxy resin lining of water mains. *Journal of the Institution of Water Engineers and Scientists* **37**, 257–271.

POYNTER, S. F. B. & MEAD, G. C. 1964 Volatile organic liquids and slime production. *Journal of Applied Bacteriology* **27**, 182–195.

SCHOENEN, D. & COLBOURNE, J. S. 1987 Microbiological evaluation of drinking water construction materials: comparison of two test procedures. *Zentralblatt für Bakteriologie und Hygiene. I.Orig.B.* In press.

SCHOENEN, D. & SCHOLER, H. F. 1985 *Drinking Water Materials: Field Observations and Methods of Investigation.* Chichester: Ellis Horwood.

SCHOFIELD, G. M. & LOCCI, R. 1984 Colonisation of components of a model hot water system by *Legionella pneumophila*. *Journal of Applied Bacteriology* **58**, 151–162.

SEAL, K. J. 1982 Biodeterioration in practice: Part 2. Case histories. *Laboratory Practice* **31**, 222–224.

STANDING COMMITTEE OF ANALYSTS (Various dates) Department of the Environment. Methods for the Examination of Waters and Associated Materials. London: Her Majesty's Stationery Office.

THORNBERRY, C., BALOWS, A., FEELEY, J. C., JAKUBOWSKI, W. (Eds.) 1984 *Legionella: Proceedings of the 2nd International Symposium*. American Society for Microbiology, Washington, D.C.

VAN DER KOOIJ, D. & HIJNEN, W. A. M. 1984 Aanwezigheid en bestrijding van *Legionella pneumophila* de veroorzaker van de veteranenzeikte in warmtapwatersystemen. H_2O **17**, 387–391.

VOSS, L., BUTTON, K. S., LORENZ, R. C. & TUOVINEN, O. H. 1986 Legionella contamination of a pre-operational water treatment plant. *Journal of the American Water Works Association*, January 1986, 70–75.

WARREN, I. 1985 The Effect of Water Quality on Certain Test Methods used to Assess Microbiological Growth on Materials in Contact with Potable Water. *Report 1059-M*. Marlow, WRc Environment.

WATKINS, I. D., TOBIN, J. O'H., DENNIS, P. J., BROWN, W., NEWNHAM, R. & KURTZ, J. B. 1985 *Legionella pneumophila* serotype 1 subgrouping by monoclonal antibodies—an epidemiological tool. *Journal of Hygiene* **95**, 211–216.

WHITE, S. F. & MAYS, G. D. 1986 *Water Supply Byelaws Guide*. Chichester: Ellis Horwood.

A Serological Method, Using a Micro ELISA Technique, for Detecting and Enumerating Sulphate-reducing Bacteria

STEFANIA BOBOWSKI* AND D. B. NEDWELL
Department of Biology, University of Essex, Colchester

Sulphate-reducing bacteria (SRB) are comprised of a group of morphologically diverse, obligately anaerobic organisms which obtain their energy by dissimilatory reduction of sulphate to sulphide. The SRB are important for a number of economic and ecological reasons (Postgate 1979). In relation to the oil industry they are found to be inhabitants of the water base of oil bearing strata and shales (Dockins *et al.* 1980), of the injection waters used in secondary oil recovery and of drilling muds used in exploration for oils. They cause corrosion of pumping machinery and storage tanks, and pollute oil products by introducing reduced sulphur into oil and its associated gas. Their economic effects are therefore of serious concern to the oil industry.

Any ecological study of SRB requires their detection and enumeration in mixed microbial communities. Traditional methods of enumeration and identification are time consuming, but immunological techniques provide one way to overcome this problem. Serological techniques previously used to attempt to identify and classify SRB include agglutination (Sefer & Pozsgi 1968; Abdollahi & Nedwell 1980), immunodiffusion (Postgate & Campbell 1963; Abdollahi & Nedwell 1980) and immunofluorescence (Smith 1982). The data obtained have been somewhat contradictory in that although some workers have reported cross-reactivity between strains or species of SRB, others have reported highly strain specific antigens.

Recently, other marker techniques have been proposed to identify SRB, including 'finger printing' by long chain fatty acids characteristic of different types of SRB (Taylor & Parkes 1983) and analysis of proteins in the outer membrane of SRB (Norqvist & Roffey 1985). However, the complexity of these techniques does not make them practicable for field application.

With the introduction of enzyme-labelled antibodies for the detection of virus antigens in tissue culture (Nakane & Pierce 1966; Wicker & Avrameas 1969), the possibility of easily detecting different strains of micro-organisms became real. Their quantitative use was described by Engvall & Perlmann (1971, 1972) who

*Present address: BP Research Centre, Sunbury on Thames, Middlesex TW16 7LN.

0–632–01793–7

claimed that ELISA techniques had sensitivities and specificities comparable to radioimmunoassay and immunofluorescence assay, but offered several advantages over them. ELISA techniques have been applied to the detection and identification of a wide variety of organisms such as *Rhizobium* (Martensson *et al.* 1984), methanogenic bacteria (Archer 1984) and xylem-linked bacteria (Saunders & French 1983).

Materials and Techniques employed in the Method

The method described here involves an indirect microplate ELISA procedure suitable for detecting SRB in both mixed cultures or natural samples. The procedure is quick and simple, efficient with respect to time, reagents and equipment, yet retains a high sensitivity.

Source of Organisms

The SRB strains used are listed in Table 1, and include the type strains of all species of *Desulfovibrio*, of *Desulfobacter postgatei* and of *Desulfotomaculum nigrificans*. Most were obtained as lyophilized cultures from the National Collection of Industrial and

TABLE 1. *Cross-reaction of polyvalent antiserum with strains of sulphate reducing bacteria, and two control non SRB strains*

Strain (antigen)		Polyvalent antiserum titre
Desulfovibrio africanus	NCIB 8401*	1600
D. gigas	NCIB 9332*	200
D. desulfuricans	NCIB 8380*	3200
D. desulfuricans	NCIB 8363	800
D. salexigens	NCIB 8403*	3200
D. vulgaris	NCIB 9442*	25600
D. vulgaris	F1	3200
(North Sea strains)	F2	1600
	F4	25600
	F8	51200
	S6	400
	S19	12800
Db. postgatei	DSM 2034*	200
Dt. nigrificans	NCIB 8351*	0
E. coli	NCTC 9002	0
S. aureus	NCTC 3750	0

D. = *Desulfovibrio*; *Db.* = *Desulfobacter*; *Dt.* = *Desulfotomaculum*.
* Type strains.

Marine Bacteria (NCIMB), Aberdeen, Scotland. Five strains of SRB isolated from the North Sea were obtained from Dr J. A. Hardy (British Petroleum Company, Sunbury on Thames, Middlesex). *Desulfobacter postgatei* (Dangast, 2ac9, DSM 2034) was obtained from the Deutsche Sammlung von Mikroorganism, Gottingen, West Germany.

Growth, and Preparation of SRB Antigens

SRB were grown in a modified Postgate's medium (Abdollahi 1979) and the cells were harvested in late exponential phase. Iron sulphide precipitate in the culture medium was first removed by centrifugation at low speed [1500 *g* for 10 min at 4°C Chillspin centrifuge (MSE Instruments, Crawley, Sussex)]. The cells were then harvested by centrifugation at 25 000 *g* for 20 min at 4°C and washed twice in phosphate-buffered saline (PBS) containing 0.15 M sodium chloride, and 50 mM phosphate buffer, pH 7.2 (Hudson & Hayes 1972). The cells were then resuspended in PBS containing 10 mM ethylene diamine tetra acetic acid (EDTA) to give a concentration of approximately 10^{10} cells/ml^{-1}, and incubated at 45°C for 2 h. Subsamples (2 ml) of cell suspension were sonicated at 0°C for 5 min (in 5 s bursts with intervening cooling in ice) at maximum frequency in a MSE Soniprep sonicator (MSE Instruments, Crawley, Sussex). Unbroken cells and debris were removed by centrifugation at 25 000 *g* for 20 min at 0°C. The supernatant was used as the antigen preparation for the ELISA procedure. The amount of protein in each antigen preparation was estimated by the method of Lowry *et al.* (1951) using bovine serum albumin as standard, and adjusted to a concentration of 1 mg/ml protein.

Antibody Production

The fifteen strains of SRB listed in Table 1 were used to raise antisera in New Zealand white rabbits (Ranch Rabbit, Crawley, Sussex).

After being bled for pre-immune serum, the animals were injected with 2 ml of a suspension of whole cells of SRB (10^{10} cells/ml in PBS) mixed with an equal volume of Freund's Complete Adjuvant. The rabbits were reinoculated 28 days and 45 days after the initial inoculation, but with the SRB cell mixed with Freund's Incomplete Adjuvant. The rabbits were bled after 52 days, and thereafter at fortnightly intervals, from the marginal ear vein. The blood was left to clot overnight at 4°C and the serum then separated and stored at −20°C. Periodic booster injections were made if the titre was low or falling. Different batches of antisera were titrated against the homologous SRB antigen preparation and standardized to give a titre of 2048. A polyvalent antiserum mixture was used in an attempt to maximize detection of as many strains or species of SRB as possible.

ELISA Procedure

This was essentially the same as that described by Engvall & Perlmann (1972) but adapted for microtitre plates. Preliminary work to examine the suitability of different plastic microtitre plates revealed that the best for the ELISA procedure were polyvinyl chloride plates with flat-bottomed wells (Sterilin Ltd, Teddington, Middlesex), as used by Voller *et al.* (1976). Preliminary analysis was carried out to determine the optimum dilutions of antigen preparation and enzyme conjugate required to give greatest colour production in the ELISA assay, while maintaining a low background of colour production with controls. A chequer board two-dimensional titration of doubling dilution of antigen against serial 1.5-fold dilutions of conjugate in the presence of an excess antibody was used. A 1:16 dilution of standard antigen preparation, and a 1:3000 dilution of conjugate were found to be the minimum concentrations giving maximum colour production and were adopted as the standard for subsequent experiments. In trial experiments the anti SRB sera were titrated against their homologous antigens, and titres between 6400 and 204 800 were obtained. All subsequent micro ELISA determinations used an antiserum dilution of 1:500, at which concentration there was a negligible colour production when pre-immune serum controls were reacted with SRB antigen preparations (i.e. a low background colour was maintained). This result contrasts with that of Singleton *et al.* (1985) who found significant colour production by *D. africanus, D. gigas* and *D. desulfuricans* in an ELISA technique using whole cells of *Desulfovibrio* spp.

Plate coating with antigen

The SRB antigen preparation in 50 mM carbonate buffer, pH 9.6 (50 μl per well) was added to each well and incubated at 37°C for 4 h and then overnight at 4°C. The plate was washed three times in 0.15 M sodium chloride solution containing 0.05% Tween 20. The plates were shaken dry between each wash.

Reaction with antiserum

Nonfractionated rabbit anti-SRB polyvalent antiserum diluted 1:500 in PBS containing 0.05% Tween 20, was added to the coated wells (50 μl per well) and incubated at 25°C for 4 h. The plates were washed as previously.

Reaction with enzyme conjugate

Goat anti-rabbit IgG conjugated to horse radish peroxidase (Bio-rad Laboratories, Richmond, California, U.S.A.), diluted 1:3000 in the same buffer as the antiserum, was added to each well (50 μl per well) and incubated at 25°C for 3 h. Comparison of the peroxidase system with that using alkaline phosphatase showed that the former was the most sensitive and was therefore adopted for routine quantitative use.

Colour development

After a further wash, a volume (50 μl) of the enzyme substrate (1 mg/ml orthophenylamine diamine in 0.2 mM phosphate–0.1 M citrate buffer, pH 5.0) was added to each well and incubated at room temperature for 1 h. The reaction was stopped by adding 10 μl sulphuric acid (2.5 M) per well.

Two sets of controls were routinely used: an antigen control in which antiserum buffer replaced antiserum; an antiserum control in which antigen buffer replaced the antigen.

Quantification

In the early experiments the reaction between antigen and antiserum was quantified by titration with doubling dilutions of the antiserum. The greatest dilution giving colour production by the micro ELISA technique above that of the background controls was regarded as the end point and the titre was the reciprocal of this dilution.

In later experiments absorbance of the colour produced in each well was directly measured in a small volume cuvette at 492 nm in a Uvicord spectrophotometer (Pye-Unicam, Cambridge). A subsample (10 μl) was removed from each well and diluted to 1 ml with buffer to give sufficient volume for the cuvette.

Estimation of SRB in Natural Samples Using ELISA

The ELISA technique was calibrated to express the colour produced (A_{492}) by the standard ELISA test in terms of an equivalent number of SRB cells. To this end known numbers of SRB cells on membrane filters were extracted, and the antigens thus obtained determined by ELISA.

Extraction procedure for SRB antigens

The ELISA technique was calibrated using three strains of SRB: *D. salexigens* (NCIB 8403), *D. desulfuricans* (NCIB 8380) and *D. vulgaris* (NCIB 9442). A 10-fold dilution series of a cell suspension of each strain was used to place different numbers of cells on the surface of 0.2 μm pore size membrane filters (2.5 cm diameter; Nuclepore, Chicago). The number of SRB cells in the original cell suspension was determined by direct microscopic counting in a Thoma counting chamber. Each filter was transferred to a thin-walled boiling tube and 2 ml of PBS containing EDTA (10 mM) was added. The tube was vortexed intermittently for 5 min to remove cells from the filter and incubated at 45°C for 2 h. The solution was then decanted, sonicated as described above and centrifuged to remove debris. The clear extract was then used as the antigen preparation in a micro ELISA determination against the polyvalent antiserum. Controls with pre-immune serum and with polyvalent antiserum against a non-SRB control organism were also set up.

The effect of the presence of large numbers of non SRB bacteria on the results of micro ELISA extraction and titration was tested by extracting filters bearing known number of cells of *D. salexigens* in the presence and absence of equal numbers of cells of *E. coli.*

In addition to cultures of SRB strains, samples of natural materials, including seawater, minewater and production water from North Sea oil rigs, were filtered and extracted as described and the SRB antigens in the extracts measured by micro ELISA. The A_{492} determined for each sample was expressed as an equivalent number of SRB cells, and compared to the Most Probable Number (MPN) count of SRB determined for the same sample. The MPN count was carried out by standard technique (Hardy, 1981) on modified API RP38 medium (Anon. 1965) with lactate as the substrate for SRB growth.

Results from the Application of the Technique

Table 1 shows the results of the examination of cross-reactivity of the polyvalent SRB antiserum against all fifteen homologous strains, together with several non SRB control organisms. The polyvalent antiserum reacted with all of the strains of SRB tested although there was some variation in titre from strain to strain. There was no cross-reaction with two non-SRB control organisms. In a recent paper Singleton *et al.* (1985) also described an ELISA technique for SRB using whole cells as the antigen, rather than antigen extracts as described here. They reported cross reactions between antisera and various heterologous strains of *Desulfovibrio.* However, these workers found considerable reaction between *D. africanus, D. gigas* and *D. desulfuricans* and pre-immune serum, which they considered to indicate the presence of lectin-like substances on their cell surface. No significant reaction with pre-immune serum was detected with any strain in the present work.

Fig. 1 shows the relationship between absorbance at 492 nm obtained with the micro ELISA technique (corrected for sample dilution) and the numbers of cells of SRB present on the membrane filters. *E. coli* controls showed no absorbance at 492 nm, and controls of pre-immune serum also gave no colour production. In general, there was close comparison for all three strains of SRB used for calibration between the value of A_{492} obtained by ELISA and the number of cells on the filter. The relationship between A_{492} and cell number was not linear. For enumeration by ELISA, an average of the A_{492} given by all three strains of SRB was used to convert A_{492} to an equivalent count of SRB cells. The technique permitted the detection of as low as 10^3 SRB cells per filter. In practice, this number of cells could be obtained by filtering smaller or larger volumes of sample.

The titre obtained in the micro ELISA technique was not affected by the presence of other, non SRB, organisms, as illustrated in Table 2. The titre remained identical in the presence or absence of *E. coli.* Similarly, other control, non SRB, organisms have been shown to have no interference with the extraction and estimation by ELISA of SRB. Results of estimations of numbers of SRB in natural

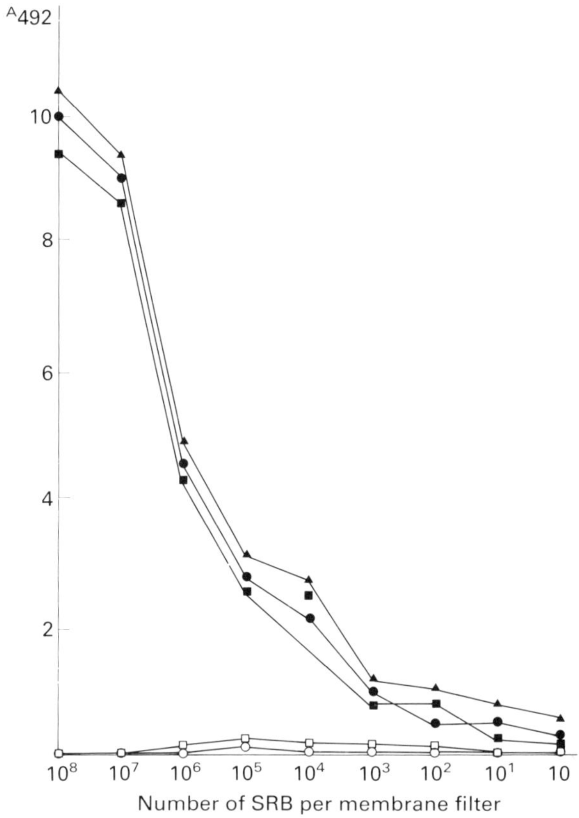

FIG. 1. A calibration graph illustrating the relationship between absorbance at 492 nm obtained with the micro ELISA technique (corrected for sample dilution) and the numbers of cells of SRB present on membrane filters. ●, Antigen NCIB 8403 *D. salexigens*; ▲, antigen NCIB 8380 *D. desulfuricans*; ■, antigen NCIB 9442 *D. vulgaris*; □, antigen NCIB 9002 *E. coli*; ○, PBS–Tween + normal serum.

samples by both ELISA enumeration, and MPN after 1 month incubation, are shown in Table 3. In general, the count by ELISA was an order of magnitude greater than that from the MPN count. This may be attributable to a number of reasons. First, the MPN counts of SRB on lactate medium will not reveal the recently-described genera (Widdel 1980) of SRB which do not metabolize lactate. Lactate-utilizers are only a small part of the total SRB community present in marine environments. The ELISA technique with an antiserum cocktail, in contrast, exhibits considerable cross reaction with all types of SRB tested, and will therefore detect and estimate a wider and more representative part of the total SRB community. This may be important in determining the overall significance of SRB numbers with regard to corrosion or souring. The counts by ELISA may in part be

TABLE 2. *Titre of SRB (10 cells per filter) by ELISA in presence and absence of equal numbers of cells of E. coli*

Antigen	Titre
D. salexigens 8403	3200
E. coli	0
D. salexigens + *E. coli*	3200

greater than MPN counts because the ELISA technique also detects the cellular antigens of non-viable SRB cells which are not detected in the MPN technique. The MPN technique detects only viable cells capable of growth in the medium provided, but SRB cells may remain antigenically intact for some time after loss of viability. The speed of breakdown of antigenic determinants after death is not known and is currently being investigated.

The time required for estimation of SRB by the ELISA technique is considerably less than that required for the MPN technique and indeed may be reduced to less than that expended in the present work. Monoclonal antibodies would facilitate the provision of a constant source of antibodies for routine use in the ELISA estimation of numbers of SRB. The regular requirement for injection of animals to produce and then standardize polyclonal antisera would thus be avoided. Raising of monoclonal antibodies to SRB is being carried out.

TABLE 3. *Comparison of estimates of SRB in representative natural samples by ELISA and MPN techniques (no. per 100 ml)*

Sample	ELISA count*	MPN count
Pond water, Esher	20	0
Hydrocarbon film on pond water	2	0
Pond water/sediment	200	20
Loch Long, BP Finnart	2	0
Askham chlorinated mains water	0.2	0
Askham day tank	0.2	0
Gainsborough stream	20	20
Gainsborough mud water supply line	20	0
North Sea water	1000	100
Production water	100	10
Mine water	100	10
Quarry water	1000	100
Stream water	100	10

* Mean of five replicates.

References

ABDOLLAHI, H. 1979 Aspects of the ecology, physiology and serology of sulphate-reducing bacteria (*Desulfovibrio*). M.Phil. thesis, University of Essex.

ABDOLLAHI, H. & NEDWELL, D. B. 1980 Serological characteristics within the genus *Desulfovibrio*. *Antonie van Leeuwenhoek, Journal of Microbiology and Serology* **56**, 73–83.

ANON. 1965 Recommended practise for biological analysis of subsurface injection waters, 2nd edn, API RP38, Dallas Division of Production, American Petroleum Institute.

ARCHER, D. B. 1984 Detection and quantification of methanogens by enzyme-linked immunosorbent assay. *Applied and Environmental Microbiology* **48**, (4), 797–801.

DOCKINS, W. S., OLSON, G. J., MCPETERS, G. A. & TURBACK, S. C. 1980 Dissimilatory bacterial sulphate reduction in Montana groundwaters. *Geomicrobiology Journal* **2**, 83–98.

ENGVALL, E. & PERLMANN, P. 1971 Enzyme-linked immunosorbant assay (ELISA) quantitative assay of IgG. *Immunochemistry* **8**, 871–874.

ENGVALL, E. & PERLMANN, P. 1972 Quantitation of specific antibodies by enzyme-linked anti-immunoglobulin in antigen-coated tubes. *Journal of Immunology* **109**, 129–135.

HARDY, J. A. 1981 The enumeration, isolation and characterisation of sulphate-reducing bacteria from North Sea waters. *Journal of Applied Bacteriology* **51**, 505–516.

HUDSON, L. & HAYES, F. C. 1972 *Practical Immunology*, 2nd edn. Oxford: Blackwell Scientific Publications.

LOWRY, O. H., ROSENBROUGH, N. J., FARR, A. L. & RANDALL, R. J. 1951 Protein measurement with the Folin phenol reagent. *Journal of Biological Chemistry* **193**, 265–275.

MARTENSSON, A. M., GUSTAFSSON, J-G & LJUNGGREN, H. D. 1984 A modified, highly sensitive enzyme-linked immunosorbent assay (ELISA) for *Rhizobium meliloti* strain identification. *Journal of General Microbiology* **130**, 247–253.

NAKANE, P. K. & PIERCE, G. B. 1966 Enzyme-labelled antibodies preparation and application for the localisation of antigens. *Journal of Histochemistry and Cytochemistry* **14**, 929–931.

NORQUIST, A. & ROFFEY, R. 1985 Biochemical and immunological study of cell envelope proteins in sulphate-reducing bacteria. *Applied and Environmental Microbiology* **50**, (1), 31–37.

POSTGATE, J. R. 1979 *The Sulphate-Reducing Bacteria*. Cambridge University Press.

POSTGATE, J. R. & CAMPBELL, L. L. 1963 Identification of Coleman's sulphate-reducing bacterium as a mesophilic relative of *Clostridium nigrificans*. *Journal of Bacteriology* **86**, 274–279.

SAUNDERS, M. S. & FRENCH, W. J. 1983 Enzyme-linked immunosorbent assay for detection of xylem-linked bacteria: use of Trinder's reagent. *Applied and Environmental Microbiology* **46**, 344–347.

SEFER, M. & POZSGI, N. 1968 Etude serologique des bacteries sulfato reductrices (genre *Desulfovibrio*) isolees en Roumainie. *Archives Roumaines de Pathologie Experimentale et de Microbiologie* **27**, 867–873.

SINGLETON, R., JR, DENIS, J. & CAMPBELL, L. L. 1985 Whole cell antigens of members of the sulphate-reducing genus *Desulfovibrio*. *Archives of Microbiology* **141**, 195–197.

SMITH, A. D. 1982 Immunofluorescence of sulphate-reducing bacteria. *Archives of Microbiology* **133**, 118–221.

TAYLOR, J. & PARKES, R. J. 1983 The cellular fatty acids of the sulphate-reducing bacteria *Desulfobacter* sp., *Desulfobulbus* sp. and *Desulfovibrio desulfuricans*. *Journal of General Microbiology* **129**, 3303–3309.

VOLLER, A., BIDWELL, D. & BARTLETT, A. 1976 In: *Immunology* eds Rose, N. R. & Freidman, H. Washington: American Society for Microbiology.

WICKER, R. & AVRAMEAS, S. 1969 Localisation of virus antigens by enzyme-labelled antibodies. *Journal of General Virology* **4**, 465–471.

WIDDEL, F. 1980 Anaeroberabbau von Fettsauren und Benzoesaure durch neu isolerte Arten sulfat-reduzierender Bakterien. Doctoral dissertation, Gottingen.

The Use of Immobilized Enzymes to Detect River Pollutants

LESLEY A. HARRISON AND K. P. FLINT
Department of Biological Sciences, University of Warwick, Coventry CV4 7AL

In Britain, at present about 30% of drinking water supplies are abstracted from lowland rivers which are subject to pollution by the accidental or deliberate discharge of wastes from a variety of industrial, agricultural and urban sources. These discharges have resulted in an increased pollution load in British rivers with an ever-increasing number of increasingly complex synthetic chemicals. The United States Environmental Protection Agency (EPA) has identified 149 'priority pollutants' which, because of their suspected toxicity or carcinogenicity, the EPA considers should not be present in finished drinking water and preferably not in river water abstracted for drinking purposes (Chapman *et al.* 1982). Consequently, it is disturbing that many of these chemicals have been identified in river water and in the finished drinking water of many U.S. cities (Cantor *et al.* 1981) and it is probably only a matter of time before this situation is repeated in British and European cities which rely on lowland river water as their primary source of drinking water.

To analyse drinking water for the presence of each of these 149 priority pollutants is both costly and time consuming as well as being impracticable, in terms of the needs of the water supply industry. Rather, there is a need to know that the river water being abstracted is 'clean', i.e. that it does not contain above the acceptable limit of any pollutant. Most of the biochemical tests or bioassays developed to screen for these pollutants are long-term tests and are unsuitable for the protection of drinking water supplies from transient pollution plugs caused by accidental discharge of a toxic material to a water course. The one exception to this is the fish cough monitor, but this has other technical problems (Cairns *et al.* 1980).

Here we are concerned with the development of an experimental system based on immobilized enzymes which can determine whether or not a river water sample is polluted with any of the priority pollutants, prior to its abstraction for use as drinking water. Any such system must be able quantitatively to determine pollution concentration within a short time (preferably 15–30 min), as in some parts of the country abstracted river water becomes available as mains drinking water within

0-632-01793-7

2 h of abstraction. Our experimental system uses immobilized phosphatase enzymes to determine the concentration of oxyanions in river water and sediment samples.

Biochemical Assays

One possible way of developing a rapid assay system is to develop a biochemical assay, based on some cellular function like respiration, enzyme activity, ATP levels or bioluminescence (Bitton 1983).

One of the most important, but least understood, actions of toxicants, like pesticides, is their effect on enzyme activity. It is rare for studies on purified, or even unpurified, enzyme preparations to include investigations into the effects on enzyme function of important environmental contaminants. However, it is known that many of these contaminants can inhibit bacterial growth and are directly toxic to protozoa or algae (Bringmann & Kuhn 1980). This toxicity must result from the impairment of some vital enzyme function, probably a respiratory or central metabolic enzyme (Iverson & Binckman 1978).

Table 1 lists some enzymes which it has been suggested could form the basis of a biochemical monitor for the presence of toxicants in water. Some enzymes are specifically inhibited by a single pollutant, e.g. the action of lead on δ-aminolaevulinic dehydratase (Goldstein 1972) but most are nonspecific and the ultimate debilitating effect on the organism is a result of a variety of nonspecific biochemical malfunctions (Christensen *et al.* 1982).

Bitton (1983) considered that biochemical tests could form the basis for

TABLE 1. *Enzymes inhibited by important environmental pollutants*

Enzyme(s)	Inhibitors	Reference
ATPase	Dieldrin Vanadate	Reidel & Christensen 1979 Bowman & Slayman 1979
Cholinesterase	Organophosphorus pesticides	Bhattacharya *et al.* 1981
Dehydrogenases	Metals Chlorinated phenols	Broecker & Zahn 1977 Liu 1981
Lipase	Parathion DDT Chlorinated phenols	Christensen *et al.* 1982
Nitrate reductase	Industrial wastes	Williamson & Johnson 1981
Oxido-reductases	Industrial wastes Metals	Busch 1982 Berkun 1982
Ribonuclease	Malathion	Christensen *et al.* 1982

bioassays which would be rapid, inexpensive and more economical in terms of space and reagents than the fish bioassays with which they would need to compete if they were to gain acceptance. However, as yet, a correlation between results from biochemical bioassays and fish bioassays is far from established. Thus, because the relationship between inhibition of enzyme activity by pollutants and the effects of these pollutants on human health is tenuous, it is unlikely that these biochemical-based toxicity tests will become standard practice in the near future. However, they are amenable to adaptation and are faster to perform and to interpret. A battery of test units, portable in form, to detect pollutants or plugs of pollutants in water close to abstraction points and linked to a microcomputer is quite feasible. The water supplier needs to know whether or not all pollutants or classes of pollutants are absent from a water, rather than the absolute concentration of a specific pollutant. Thus non-specific enzyme tests, able to respond to a wide range of pollutants, have their place, and, with the development of immobilization techniques which allow enzymes to be reused and the development of increasingly miniaturized enzyme-based biosensors, the future for enzyme bioassays looks promising. However, there are two major problems to be overcome: first, to establish enzyme stability under operating conditions (i.e. in the field) and second, the need for detection of sub-mg/l concentrations of most of the priority pollutants, amounts which cause little, if any, inhibition of enzyme activity. Hence there is a need either for enzymes which are very sensitive to the given pollutants, or for very sensitive assay procedures. The latter could require the concentration of pollutants from water samples prior to testing.

Methods for the Immobilization of Enzymes

Once a suitable enzyme has been identified as being of use for the detection of pollutants, it must be immobilized if it is to be reusable. There are a wide range of disparate immobilization techniques ranging from relatively simple processes, such as precipitation with bifunctional agents, to complex multistep processes involving the covalent bonding of the enzyme to modified organic and inorganic surfaces or entrapment in a polymeric gel. Over the last decade or so, more than one hundred different immobilization methods have been described in the literature; these have been reviewed by Klibanov (1983).

Biosensors for the Analysis of Environmental Samples

Biosensors are analytical instruments which consist of biologically derived catalysts like enzymes, tissues, organelles or whole cells, usually in an immobilized form, linked to a physico-chemical sensing device, like an ion-specific electrode. These biosensors can be very specific, as the enzymes used can themselves be very specific. Enzyme reactions can be monitored with comparative ease using colorimetric, spectrophotometric or amperometric methods. This allows the adaptation of existing equipment for the detection of chemicals, e.g. autoanalysers with immobilized enzymes have been developed for the quantitative determination of specific substances in environmental samples (Sundaram *et al.* 1979). The use of immobilized enzymes for assays could reduce costs, particularly in the clinical and environmental areas, where the routine evaluation of a great many samples is required.

To date biosensors have been developed primarily for clinical use, but industrial applications of biosensors, particularly in the control of fermentation processes, have also been prominent. However, few reports mention the potential of these instruments for the environmental monitoring of trace amounts of substrates or, more importantly, of pollutants. Sternberg *et al.* (1980) discussed the use of a detector of trace amounts of glucose for environmental analyses and Neujhar (1984) has reviewed some possible applications of biosensors, including their use in environmental areas.

Most of the biosensors so far designed and described are for the detection of specific substances which are almost invariably the substrate for the immobilized enzyme. It is highly unlikely that typical aquatic pollutants would be the substrates for an enzyme, and if biosensors are to be used for the detection of pollutants they will have to be constructed such that it is the inhibition of an enzyme reaction which is being monitored. This change in emphasis will require rethinking of the basic enzyme electrode/biosensor design and construction, as well as changes in their mode of operation.

There are two types of immobilized enzyme systems which could be used for the detection of environmental pollutants. In the first type, a bioprobe, the enzyme is immobilized in close contact with the sensor, usually fixed covalently onto a membrane and the electrode surface. Guilbault (1983) described a typical enzyme electrode which could be used for the detection of glucose. This incorporated a platinum disc embedded in the end of a glass cylindrical insulated support with a platinum wire connecting the disc with the detection instrument. The enzyme, glucose oxidase, was covalently fixed to a collagen membrane or to the surface of preformed polyacrylamide beads which were then held by a cellophane membrane in contact with the electrode.

An alternative system is the bioreactor based on the autoanalyser. Here, the enzyme is attached by covalent bonding to the inside of a nylon tube or entrapped in a polymeric matrix which is then inserted in the autoanalyser. This type of instrument has been used to determine the concentration of glucose, urea, creatinine, glycerol, pyruvate and citrulline in clinical specimens (Sundaram *et al.* 1979; Guilbault 1983). This type of instrument would probably be of more importance in the monitoring for trace environmental pollutants. One possible mode of operation would be to pass the suspect water sample through the tube containing the immobilized enzyme for a fixed period of time and then to assay for the residual enzyme activity by replacing the water stream with a reaction mixture of substrate plus buffer. The pollutants/inhibitors should react with the enzyme causing inhibition, and by knowing the time for which the water stream was passed and the volume of water which passed through the column an average pollution load could be calculated.

The Detection of Environmental Pollutants using Immobilized Enzymes

The use of biosensors, incorporating immobilized enzymes for the detection of environmental contaminants, particularly the priority pollutants, has rarely been reported. There are ion-specific electrodes and biosensors for the detection of nitrate, nitrite, phosphate and sulphate, but these are rarely important pollutants. Organophosphorus and carbamate pesticides have been detected at sub-ppm concentrations using immobilized acetylcholine–esterase (Bhattacharya *et al.* 1981; Goodson & Jacobs 1976) and phenol concentrations have been determined quantitatively using immobilized phenol oxygenase (Macholen & Schanel 1977). Some metal species can also be determined using enzymatic means, e.g. magnesium ions with alkaline phosphatase apoenzyme and calcium ions with tyrosinase apoenzyme (Matiasson *et al.* 1979).

The Use of Phosphatase Enzymes to Detect Oxyanions in River Water

To illustrate how immobilized enzymes could be used in environmental studies, a neutral phosphatase enzyme has been used in a model system for the detection of oxyanions, particularly vanadate, in river water. Neutral phosphatase has been used because it is a stable enzyme and easily assayed using the colorimetric substrate, p-nitrophenyl phosphate (pNPP). It can be easily immobilized to surfaces and entrapped into polymeric gel matrices. Almost all of the work described here has been undertaken with polyacrylamide gel-immobilized preparations but equally good results have been obtained using calcium alginate-immobilized samples or preparations covalently bonded to cellulose acetate membranes using glutaraldehyde.

Vanadium occurs as vanadate in waters although the oxidation state is dependent on ionic concentration and on pH (Pope & Dale 1968). It is acutely toxic to fish (Stendahl & Sprague 1982), lichens (Le Sueur & Puckett 1980) and plants (Saxe & Rajagopal 1981). The mechanism of its toxicity is believed to be via inhibition of phosphatase or ATPase function (Bowman & Slayman 1979; Wu & Phillis 1979).

Vanadate is a common constituent of crude oils, and it is possible that vanadate could be used as an indicator or a tracer of oil pollution in marine or fresh waters.

Materials and Methods

Organism

A *Pseudomonas fluorescens* strain isolated from a local stream has been shown to possess a constitutive neutral phosphatase enzyme with a pH optimum of 7.0. The organism was routinely cultured in nutrient broth at 30°C until stationary phase was reached.

Assay of phosphatase activity

0.1 ml of cell-free extract, or a single weighed piece of gel, was added to 3 ml of 0.1 M tris-tes-acetic acid (TTA) buffer pH 7.0 (Flint & Hopton 1976). After 10 min preincubation at 30°C, 0.1 ml of 5% (w/v) pNPP in distilled water was added and the reaction allowed to proceed for 30 min. The reaction was stopped by the addition of 0.1 ml 10 M NaOH. The increase in absorption at 420 nm was read on a spectrophotometer in 1 cm glass cuvettes. Controls containing no enzyme and blanks with no substrate were also included. Enzyme activity was calculated as μM pNP released/unit time (/unit of gel weight where necessary) using a standard curve for the absorption of pNP at 420 nm.

For the pH/activity profiles, TTA buffer of the appropriate pH was used in place of the pH 7.0 buffer. For the inhibition studies, 0.3 ml of the appropriate inhibitor in TTA buffer was added to 2.7 ml of buffer.

Preparation of enzyme extracts

Stationary phase cultures were harvested and the cells resuspended in one-tenth volume of sterile distilled water. Enzyme extracts were prepared by ultrasonication, with cooling in ice, until >90% of the cells were broken as determined by microscopy. The cell debris was removed by centrifugation and the crude extracts dialysed overnight at 4°C in 0.1 M TTA, pH 7.0. This crude extract was used without further purification, although it was diluted such that 0.1 ml used in the above assay gave an absorption change of approximately 1.0 after 30 min incubation.

Immobilization in polyacrylamide gel

The crude extract was immobilized into polyacrylamide gel as follows. The following constituents were mixed in a beaker: 6.75 ml acrylamide stock solution (2.2 g acrylamide and 0.8 g N,N′-methylene-bis-acrylamide in 100 ml distilled water), 15 ml enzyme extract, 1.5 ml aqueous ammonium persulphate (1.5 mg/l) and 40 µl N,N,N′,N′-tetramethylethylenediamine (TEMED). The mixture was poured into petri dishes and allowed to polymerize at room temperature. The

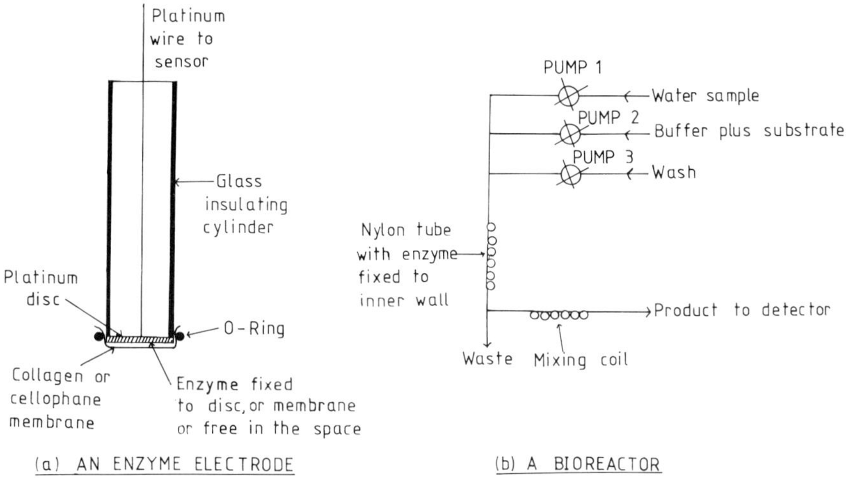

FIG. 1. (a) A possible mode of construction of an enzyme electrode, based on the enzyme being held in contact with a platinum electrode by a membrane. (b) A bioreactor. The enzyme is covalently bonded to the inner wall of a nylon tube. Alternatively, this tube could be replaced by a column containing a gel-immobilized enzyme. Pumps 1, 2 and 3 operate at different times pumping sample, reactants and wash through the immobilized enzyme preparation.

polymerized gel was manually cut into pieces, washed repeatedly in 0.1 M TTA buffer and finally stored at 4°C in a small amount of buffer.

Properties of the immobilized enzyme

Phosphatase assays were all carried out at pH 7.0 and 30°C. The reaction was tested for linearity with increasing weight of a single piece of gel (Fig. 2). The reaction was linear up to a weight of 100 mg. Above this size the reaction ceased to be linear presumably due to diffusion limitation. Enzyme in the centre of the gel would be unreactive as no substrate would penetrate that far. By using longer thinner pieces of gel the reaction could probably have been made linear for a larger piece of gel. This demonstrates how important the shape of the gel piece is to the reaction. The greater the surface area available for diffusion the higher the reaction rate will be for the same amount of enzyme. We were able to demonstrate that enzyme entrapment in the gel was very effective: a 50 mg piece of gel equivalent to approximately 25 μl of initial protein solution gave an absorbance reading of 0.95 for the reaction whereas the free enzyme gave a reading of 1.25 for the same incubation time. All subsequent experiments were conducted using gel pieces of approximately 50 mg weight and of the same size and shape. Corrections were made in the calculation of enzyme activity for the difference in gel weight used, i.e. corrected for the different enzyme concentration present in each sample. We have assumed a homogeneous distribution of enzyme in the gel.

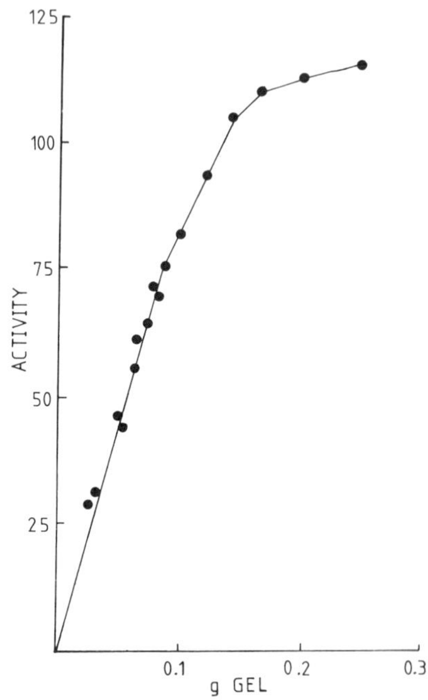

FIG. 2. Increase in activity with increasing size of a single piece of polyacrylamide gel containing immobilized neutral phosphatase enzyme. Incubation was at 30°C for 30 min in 0.1 M TTA buffer, pH 7.0. Activity was expressed as μM pNP released/ml/h.

Linearity of reaction and stability

The reaction was shown to be linear with time for up to 2 h (Fig. 3). This was taken to indicate that there was no product inhibition caused by build up of product in the gel piece due to poor diffusion from the gel. The enzymes were stable for at least 12 h at 30°C and for more than 6 months at 4°C. Pieces stored at 4°C were used daily, washed, then stored again and pieces stored over a 4 month experimental period completely retained their activity. This shows that immobilized enzymes could be reused for a considerable time providing that they remained stable, that there was no irreversible inhibition and that the gel retained its physical and mechanical stability. Polyacrylamide is the favoured entrapment matrix as, unlike alginate, it is not susceptible to microbial attack. Alginate beads used in the same way gradually lost activity after 1 week, probably due to the deterioration of bead structure brought about by the removal of calcium ions during phosphatase assays (phosphate released from pNPP by enzyme activity can chelate calcium ions forming the bridges which give the alginate its polymeric structure). This problem could probably be alleviated by storage in calcium solutions.

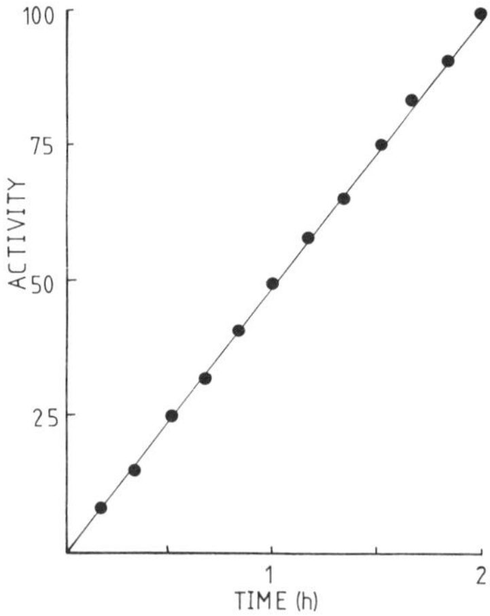

FIG. 3. Release of pNP (μM/ml/h) with time (hours) for a 50 mg piece of polyacrylamide gel containing immobilized neutral phosphatase enzyme. Incubation was at 30°C and pH 7.0 in 0.1 M TTA buffer.

pH/activity profile

Free and immobilized preparations showed no change in the optimum pH for enzyme activity but the immobilized enzyme had a higher activity in each tail of the profile (Fig. 4). This is not an uncommon phenomenon with immobilized enzymes. With a covalently bonded enzyme, it is often the pH optimum which can be altered due to restricted movement of the tertiary structure around the active site of the enzyme. With an entrapped enzyme, a change in the activity in the tails of the pH/activity profile is more commonly observed. The enzyme active site is unlikely to undergo any steric hindrance in a polymeric gel matrix. Therefore any increase in activity is probably artefactual and due to selective buffering of the gel. The pH of the bulk phase may not necessarily be the same as that in the immediate vicinity of the enzyme, i.e. micro-environments are created within the gel pores and this causes an apparent increase in activity at the pH extremes whereas in fact it is the pH which is different.

Inhibition by oxyanions

Inhibition of enzyme activity and the potential of the immobilized preparations for use as a monitor of oxyanion pollutants was examined using known competitive inhibitors of neutral phosphatase (phosphate, tungstate, vanadate and molybdate). Some metal ions were also examined. Table 2 shows the concentration of inhibitors which caused 50% inhibition of phosphatase activity with pNPP as the substrate. The most inhibitory oxyanions were vanadate and tungstate, causing 50% inhibition of enzyme activity at 0.51 ppm and 0.92 ppm respectively. There was little

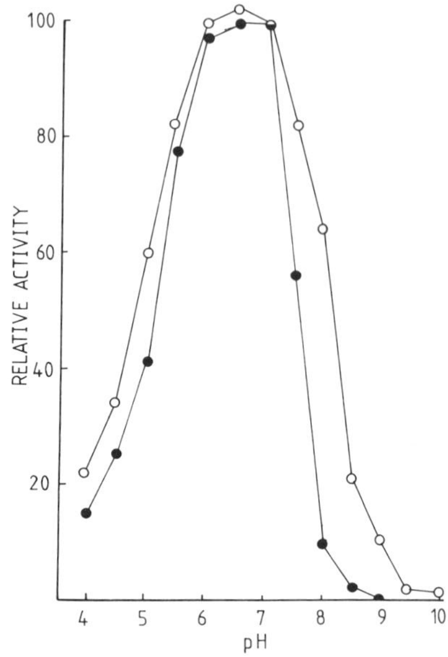

FIG. 4. pH/activity profiles of the free (●) and polyacrylamide gel-immobilized neutral phosphatase enzyme (○). Incubation was at 30°C for 30 min in 0.1 M TTA buffer at the appropriate pH. All activities are expressed relative to that at pH 7.0.

difference in inhibition of the free and immobilized enzymes. Only copper, of the metal ions tested, showed much inhibition.

The inhibition caused by vanadate and tungstate could be differentiated by adjusting the pH of the system to 8.5, adding 10 mM EDTA and leaving for 15 min. Removal of the gel pieces and return to pH 7.0 and reassay showed that vanadate

TABLE 2. *Effects of inhibitors on activity of free and immobilized neutral phosphatase preparations*

	Concentration causing 50% inhibition of activity	
Inhibitor	Free enzyme (mg/l)	Immobilized enzyme (mg/l)
Phosphate	2170 (70 mM)	3100 (100 mM)
Vanadate	0.51 (10 μM)	0.51 (10 μM)
Tungstate	0.37 (2 μM)	0.92 (5 μM)
Molybdate	2.88 (30 μM)	3.84 (40 μM)
Zinc	196 (3 mM)	458 (7 mM)
Copper	6.4 (100 μM)	19.1 (300 μM)
Nickel	176 (3 mM)	235 (4 mM)

inhibition was reversible whereas that caused by tungstate and the other oxyanions was not.

Inhibition by environmental samples

Water and sediment samples were collected from numerous sites along the Rivers Sowe and Sherbourne in Coventry. Table 3 shows the effects of these samples on phosphatase activity of the immobilized preparations. The sediment samples were treated as follows. 1 g of sediment was shaken with 10 ml TTA buffer for 12 h and the supernatant fluid decanted, filter sterilized and used. The only samples causing inhibition of the enzyme were the sediment samples collected from close to the discharge point for Coventry sewage works, and the site of an old metal working factory (Renolds Chains). The only water sample causing inhibition was one known to have been polluted by paraffin oil, a source of vanadate. No attempts to find the identity of the inhibitor were made. It is our premise that one does not need to know its identity, merely that it is present.

So far this enzyme has only been used in a batch system where gel pieces are incubated in a vial with the water sample. It is possible that,using a flow system similar to those illustrated in Fig. 1, vanadate and tungstate could be detected at much lower concentrations. It should be possible to detect at least 0.1 ppm of vanadium which is the WHO limit for drinking water and certainly less than the LC_{50} for rainbow trout (Stendahl & Sprague 1982). This is by far the most rapid and simple method for the detection of vanadium as an oxyanion but unfortunately it is not specific for vanadium, although vanadate inhibition can be differentiated from inhibition caused by the other oxyanions.

Phosphate, which is the commonest oxyanion in water, is unlikely to cause any inhibition of this enzyme at natural concentrations. The usual phosphate concentration of these rivers is less than 10 mg/l, whereas in excess of 3000 mg/l phosphate was required to bring about 50% inhibition of enzyme activity. This would not be the case with alkaline phosphatases which are inhibited by substantially

TABLE 3. *Inhibition of polyacrylamide gel-immobilized neutral phosphatase preparations by river waters and sediments*

Sample Site (Ordnance Survey Reference)	*Control activity* (%)
River Sowe, spring source (SP 866 291)	101
River Sherbourne, Renolds Chains (SP 794 315)	100
River Sowe, below sewage discharge (SP 734 732)	96
River Sowe, paraffin oil pollution (SP 741 334)	73
River Sowe, sediment, source (SP 866 291)	99
River Sowe, sediment, Renolds Chains (SP 794 315)	61
River Sowe, sediment, below sewage discharge (SP 734 332)	72

lower concentrations, e.g. 50% inhibition of the derepressible alkaline phosphatase of *Escherichia coli* is caused by less than 1 mg/l phosphate.

Future Prospects

Enzyme assays offer a relatively cheap method for analysing water samples for pollutants. The advantages they offer over traditional fish assays or other bioassays are: (1) reproductibility, providing that a standard, or purified, enzyme preparation is used; (2) a specific or a broad range of inhibitors can be identified, depending upon the enzyme(s) selected; (3) versatility, afforded by the multitude of immobilization techniques and sensing equipment available; (4) maintenance costs reduced (fish need feeding and man-hours of handling; enzymes can be left in an unattended module).

However, there is a need to find enzymes which are inhibited by sub-ppm concentrations of organic compounds, especially the chlorinated pesticides and phenols. One possible advance in this direction could arise from the developing field which links electronics and biological catalysts. A sensitive bioactivity monitor could be developed using a reference electrode, a mediator to facilitate electron transfer to a working electrode and a power supply (Turner *et al.* 1983). This technique could be used to assess microbial contamination of milk and water or to assay the concentrations of substrates, inhibitors or activators of the enzyme biocatalysts. This type of sensor will almost certainly be based on the oxido-reductase group of enzymes, which make up the bulk of enzymes used in biosensors at present (Aston & Turner 1984). The better performance characteristics of these direct enzyme-based sensors in the laboratory are due to the tight coupling between the redox reaction and the electrode, especially with the use of electron transfer mediators like dimethyl ferrocene. Aston & Turner (1984) have recently described a methanol sensor based on immobilized methanol dehydrogenase able to detect less than 0.001 ppm methanol, far below the level of detection of the conventional sensors, and a glucose detector using NAD-independent glucose dehydrogenase immobilized on a modified carbon electrode. This latter enzyme requires no added cofactors, but the assay must be carried out in an anaerobic environment. There is a vast range of oxidoreductase enzymes available commercially and also present in cells but not yet commercially purified, with, as yet, untapped potential for the development of more enzyme electrodes. As our expertise in this field increases these electrodes will become more miniaturized, computerized and more versatile, such that they will be used not only to measure substrates as at present but also to measure inhibitors.

Environmental monitoring is, at present, a relatively untouched field as it is not viewed as giving such a good return for investment as the clinical and industrial (particularly fine chemicals) fields. Water authorities are concerned to maintain the high quality of our water supplies as well as to protect water courses from pollution.

They need to know that water being abstracted from a lowland river is safe for conversion to drinking water. This requires a rapid bioassay, as in some parts of the country water is available in the mains only 2–3 h after abstraction. Traditional bioassays do not meet this need and, although the fish cough monitor does, it is expensive to use and maintain. An ideal replacement would be an enzyme-based sensor, constructed on a modular system in which a bank of enzymes could be used to detect pollutants and by microcomputer operations switch on and off abstraction pumps. However, this aim is some years in the future, but current advances in biotechnology must bring the day nearer.

References

ASTON, W. J. & TURNER, A. P. F. 1984 Biosensors and biofuel cells. *Biotechnology and Genetic Engineering Reviews* **1**, 90–120.

BERKUN, M. 1982 Method for the estimation of BOD parameter. II. Effects of inorganic metal toxicity on BOD. *Water Research* **16**, 559–564.

BHATTACHARYA, S., ALSEN, C., KRUSE, H. & VALENTIN, P. 1981 Detection of organophosphate insecticide by an immobilised enzyme system. *Environmental Science and Technology* **15**, 1352–1355.

BITTON, G. 1983 Bacterial and biochemical tests for assessing chemical toxicity in the aquatic environment: a review. *CRC Critical Reviews in Environmental Control* **13**, 51–67.

BOWMAN, B. J. & SLAYMAN, C. W. 1979 The effects of vanadate on the plasma membrane ATPase of *Neurospora crassa*. *Journal of Biological Chemistry* **254**, 2928–2934.

BRINGMANN, G. & KUHN, R. 1980 Comparison of the toxicity thresholds of water pollutants to bacteria, algae and protozoa in the cell multiplication inhibition test. *Water Research* **14**, 213–241.

BROECKER, B. & ZAHN, R. 1977 The performance of activated sludge plants compared with the results of various bacterial toxicity tests: a study with 3,5-dichlorophenol. *Water Research* **11**, 165–172.

BUSCH, A. W. 1982 Bioassay technique for relative toxicity in water pollution control. *Journal of the Water Pollution Control Federation* **54**, 1152–1154.

CAIRNS, J. J., THOMPSON, K. W., LANDERS, J. D., MCKEE, M. J. & HENDRICKS, A. C. 1980 Suitability of some fresh water and marine fishes for use with a mini-computer interfaced biological monitoring system. *Water Resources Bulletin* **16**, 421–427.

CANTOR, K. P., KOPFLER, F. C., HOOVER, R. N. & STRASSER, P. H. 1981 Cancer epidemiology as related to chemicals in drinking water. *Toxicological and Environmental Chemistry Reviews* **4**, 47–65.

CHAPMAN, P. M., ROMBERG, G. P. & VIGERS, G. A. 1982 Design of monitoring studies for priority pollutants. *Journal of the Water Pollution Control Federation* **54**, 292–297.

CHRISTENSEN, G. M., OLSON, D. & RIEDEL, B. 1982 Chemical effects on the activity of eight enzymes: a review and a discussion relevant to environmental monitoring. *Environmental Research* **29**, 247–255.

FLINT, K. P. & HOPTON, J. W. 1976 Some properties of a neutral phosphatase of a bacterium isolated from a trickling filter effluent. *European Journal of Applied Microbiology* **3**, 237–243.

GOLDSTEIN, G. 1972 Biochemical indicators of environmental pollution. *Indicators of Environmental Quality*, ed. Thomas, W. A., pp. 113–122. New York: Plenum Press.

GOODSON, L. H. & JACOBS, W. B. 1976 Monitoring of air and water for enzyme inhibitors. *Methods in Enzymology* **44**, 647–660.

GUILBAULT, G. G. 1983 Immobilised enzyme electrode probes. *Solid Phase Biochemistry: Analytical and Synthetic Aspects*, ed. Scouten, W. H., pp. 479–505. New York: Wiley & Sons.

IVERSON, W. P. & BINCKMAN, F. E. 1978 Microbial metabolism of heavy metals. *Water Pollution Microbiology*, Vol. II, ed. Mitchell, R., pp. 201–232. New York: Wiley & Sons.

KLIBANOV, A. M. 1983 Immobilised enzymes and cells as practical catalysts. *Science* **219**, 722–727.

LE SUEUR, P. & PUCKETT, K. J. 1980 Effect of vanadium on the phosphatase activity of lichens. *Canadian Journal of Botany* **58**, 502–504.

LIU, D. 1981 A rapid biochemical test for measuring chemical toxicity. *Bulletin for Environmental Contamination and Toxicology* **26**, 145–149.

MACHOLEN, L. & SCHANEL, L. 1977 Enzyme electrode with immobilised polyphenol oxidase for determination of phenolic substrates. *Collection of Czechoslovak Chemical Communications* **42**, 3667–3575.

MATIASSON, B., NILSSON, H. & OLSSON, B. 1979 An apoenzyme electrode. *Journal of Applied Biochemistry* **1**, 377–384.

NEUJAHR, H. Y. 1984 Biosensors for environmental control. *Biotechnology and Genetic Engineering Reviews* **1**, 167–186.

POPE, M. T. & DALE, B. W. 1968 Isopoly-vanadates, -niobates, and -tantalates. *Quarterly Review of the Chemical Society* **22**, 527–548.

REIDEL, B. & CHRISTENSEN, G. 1979 Effect of selected water toxicants and other chemicals upon adenosine triphosphatase activity. *Bulletin for Environmental Contamination and Toxicology* **23**, 365–368.

SAXE, H. & RAJAGOPAL, R. 1981 Effect of vanadate on bean leaf movement, stomatal conductance, barley leaf unrolling, respiration and phosphatase activity. *Plant Physiology* **68**, 880–884.

STENDAHL, D. H. & SPRAGUE, J. B. 1982 Effects of water hardness and pH on vanadium lethality to rainbow trout. *Water Research* **16**, 1479–1488.

STERNBERG, T., APOTEKER, A. & THEVENO, C. D. R. 1980 Trace glucose electrode for clinical, food and environmental determinations. *Analytical Chemistry, Symposia Series*, **2**, 461–473.

SUNDARAM, P. V., BLUMENBERG, B. & HINSCH, W. 1979 Routine glucose determination in serum by use of an immobilised glucose dehydrogenase nylon-tube reactor. *Clinical Chemistry* **25**, 1436–1439.

TURNER, A. P. F., RAMSAY, G. & HIGGINS, I. J. 1983 Applications of electron transfer between biological systems and electrodes. *Biochemical Society Transactions* **11**, 381–384.

WILLIAMSON, K. J. & JOHNSON, D. G. 1981 A bacterial bioassay for assessment of waste water quality. *Water Research* **15**, 383–390.

WU, P. H. & PHILLIS, J. W. 1979 Effects of vanadate on brain (Na^+-K^+)ATPase and p-nitrophenylphosphatase interaction with mono- and divalent ions and with noradrenaline. *International Journal of Biochemistry* **10**, 629–635.

Spatial Distribution of Spoilage Microbes in Fluids using a Gel System

E. C. HILL, G. C. HILL and D. A. ROBBINS
E. C. Hill & Associates, Unit M22, Cardiff Workshops, East Moors, Cardiff CF1 5EG

Anaerobic sulphide generation by micro-organisms always occurs due to a consortium of organisms. When it occurs in fluids (e.g. aqueous cutting oils, fuel tank water bottoms, bilge water) the organisms can be demonstrated in conventional samples using conventional culture methods. However, in the fluid itself there are marked gradients of nutrients, oxygen, pH, Eh, etc., and the microbes present are spatially organized in response to these gradients. This is not apparent from conventional sampling and testing and the real extent of sulphide generation is easily missed.

The foul smell of sulphide from metal-working fluids is a well-recognized phenomenon. It is symptomatic of stagnant systems and can be attributed to microbial activity. Although the sulphide-generating bacteria (SGB) are frequently referred to as sulphate reducing bacteria (SRB) it has been demonstrated by Abu Shaqra (1983) that a variety of sulphur-containing components of metal-working fluids, particularly petroleum sulphonates, organic polysulphides and sulphurized oils and fats can all be degraded to yield sulphide, although the susceptibility of different molecular configurations varies. The hydrogen sulphide gas produced is not only unpleasant but also extremely toxic (more toxic than hydrogen cyanide) and it can and does kill. Because it progressively anaesthetizes our sense organs it may seem to disappear when in fact it is becoming more dangerous. The threshold limit value in U.K. is 10 mg/m^3. Sulphide can also be a skin irritant. It stains and corrodes ferrous and non-ferrous metals and discolours cutting fluids. If emulsifying chemicals degrade to yield sulphide the emulsion becomes unstable.

Abu Shaqra (1983) and Abu Shaqra & Hill (1986) have described a technique for demonstrating the distribution of SGB in stagnant metal cutting fluid based on the technique of Wimpenny *et al.* (1981). An inoculated fluid in a glass beaker is gelled by the addition of a little agar and incubated at 30°C. Gradients of oxygen, Eh, pH, and nutrients become established in the gel and microbes distribute spatially in response to these gradients. A steel coupon in the bottom of the gel

Industrial
Microbiological Testing

0-632-01793-7

promotes a negative potential there; this is essential for significant sulphide generation. Appropriate probes and core samples then reveal the true ecology of the system under investigation.

By observation of the production of black sulphide in the gel it can be seen that SGB proliferate in the vicinity of the steel coupon. They never flourish in the absence of the aerobic (or facultative) degradative bacteria as these bacteria supply essential nutrient organic acids to SGB by degrading cutting oil components and they also remove free oxygen and reduce the Eh. Hence their presence before and during sulphide generation is essential, although the anaerobic bacteria, *Desulfovibrio* or *Desulfotomaculum*, are needed to actually generate sulphide from the sulphur-containing components. The molecular structure of these may have been altered by this time by degradative bacteria.

The technique has proved useful in investigating the comparative degradability of different fluid formulations as it proves a good model of a fluid sump where pH, Eh, oxygen and nutrient gradients occur. The gel system also gives a dramatic visual assessment of the spatial distribution of sulphide generation in fluids. (See Fig. 1.)

The Method

One method of setting up gels is as follows:

(1) 100 ml of metal-working fluid is made up in a beaker by adding the

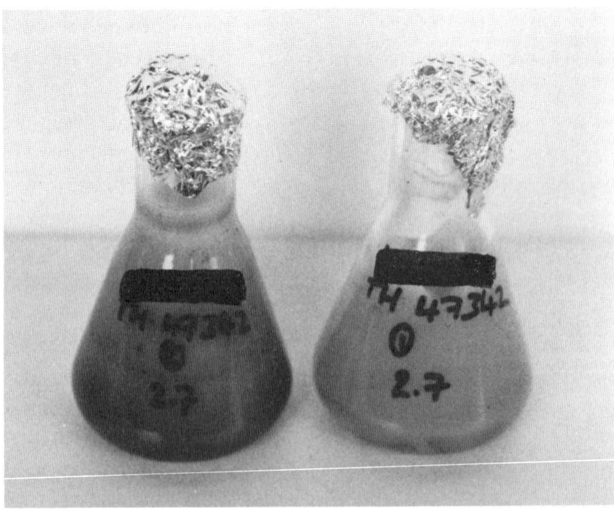

FIG. 1. Metal-working fluid gels. An infected metal-working fluid showing black sulphide generation. After 14 days (left) and 7 days (right).

concentrate to tap water at double the concentration to be tested; e.g. to test a cutting emulsion at 3% make up a 6% dispersion in tap water.

(2) 100 ml of gelling agar is made up by adding 10 mg/l agar (e.g. Bacto agar, Difco Laboratories) to water and bringing to the boil to dissolve.

(3) Add 0.1 ml of an appropriate test inoculum to the cutting fluid (e.g. a spoiled field sample). We have found that it is best if possible to use an adapted inoculum. This can be prepared by contaminating a dilute sample of the fluid under test in a large flask with various spoiled samples plus marine mud known to contain SGB. Addition of metal swarf to this flask accelerates the establishment of heavy SGB infection. Once a heavy infection is established, this fluid can be used to inoculate the gel.

(4) Drop a polished steel coupon in the bottom of the beaker containing the metal-working fluid.

(5) Add the gelling agar cooled to approximately 45°C to the fluid and stir.

(6) The gel is allowed to set and then incubated at the temperature at which spoilage is likely to occur.

Examination of the Gel

Sulphide generation is signified by a blackening of the gel, particularly at the bottom of the beaker. The rapidity and extent to which this occurs is an indication of the susceptibility of the product to SGB spoilage.

After a specified time interval the gel may be sacrificed for analysis. By setting up several identical gels, beakers can be sacrificed after different time intervals.

A sterile cork borer can be used to take a sample through the gel and portions of the gel at varying depths counted for heterotrophic bacteria in Tryptose Blood Agar Base (Oxoid CM233), SGB in Iron Sulphite Agar (Oxoid CM79) and yeasts and moulds on Malt Extract Agar (Difco 0112-17-16).

The polished steel coupon can be examined for black sulphide deposits and pitting corrosion.

Eh and pH probes can also be used in the gel.

Detection of Sulphide-Generating Bacteria (SGB)

A development of the gel technique can be used to detect simply and rapidly a sulphide generating consortium of microbes (Hill 1984). A tube containing *c.* 10 ml of a growth medium for SGB is prepared as a porous gel. A large sample (*c.* 2 ml) of the fluid to be tested is poured on to the gel and the tube is incubated at 30°C. Because of the large size of the inoculum the tube is rapidly denuded of oxygen and becomes anaerobic and a negative Eh is established. Microbes migrate into the gel and if SGB are present a black line forms at the gel interface, sometimes overnight, and expands into the gel with time to form a black zone (see Fig. 2). The speed at which this happens indicates semi-quantitatively the numbers of SGB present. This test, one of a range of Sig Tests (Significance Tests) is now available commercially (Echa Products, Cardiff Workshops, East Moors, Cardiff).

Discussion

Because conventional sampling techniques rarely specify the precise location at which samples are taken, it is difficult to appreciate the spatial distribution of organisms in a large fluid system. For example, oxygen may be depleted within a few millimetres of the surface of a heavily infected metal-working fluid. Invariably there is a completely anaerobic region at the bottom of the sump. By stabilizing gradients a good laboratory model of a stagnant fluid system is provided. Only a small volume of sample is needed; gels can successfully be made up as 20 ml

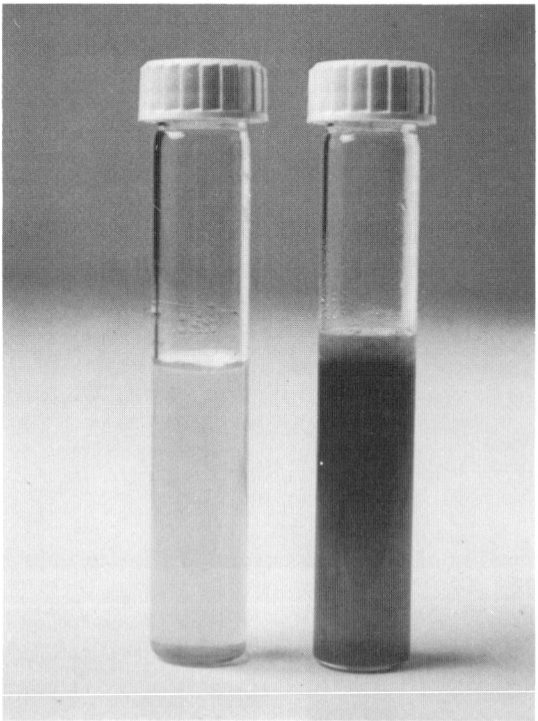

FIG. 2. Sig sulphide field test. Unused tube on left; strong positive on right after addition of sample and incubation for 3 days at 30°C.

volumes in universal bottles. In conventional laboratory tests using fluids in flasks circulation of fluid by convection currents does not allow the establishment of the marked Eh, pH, oxygen and nutrient gradients found in industrial fluid sumps and in the gel system.

References

ABU SHAQRA, Q. M. 1983 Microbial generation of sulphide from aqueous cutting fluids. Ph.D. thesis, University of Wales.

ABU SHAQRA, Q. M. & HILL, E. C. 1986 A new approach to the ecology and corrosivity of infected metal-working fluids using a gel stabilised oil emulsion model system. *Biodeterioration 6*, eds Barry, S. & Houghton, D. R., pp. 440–447. Slough, U.K.: C.A.B. International.

HILL, E. C. 1984 Micro-organisms—numbers, types, significance, detection. *Monitoring and Maintenance of Aqueous Metal-working Fluids*, eds Chater, K. W. A. & Hill, E. C., pp. 97–112. Chichester: John Wiley & Sons.

WIMPENNY, J. W. T., COOMBS, J. P., LOVITT, R. W. & WHITTAKER, S. G. 1981 A gel-stabilised model eco-system for investigating microbial growth in spatially ordered solute gradients. *Journal of General Microbiology* **127**, 277–287.

Application of Gel-stabilized Model Systems to the Study of Microbial Communities Involved in the Anaerobic Decomposition of Oil

ELAINE A. REES, CAROLINE S. TUGHAN, MOYA A. RUSS AND R. A. HERBERT

Department of Biological Sciences, University of Dundee, Dundee DD1 4HN

Whilst the economic significance and consequence of microbial activity has long been appreciated and exploited by man, it is only in recent years that he has begun to comprehend the fundamental ecological and physiological principles underlying many of these processes. There are several reasons why advances in microbial ecology have lagged behind those in other areas of microbiology. The most important of these have been the absence of appropriate methods to study highly complex and often spatially organized microbial ecosystems, and secondly that microbiologists have traditionally been conditioned to think only in terms of single pure cultures whereas most natural environments contain a great diversity of physiological groups. Whilst studies using pure cultures have led to great advances in microbial physiology, there is now an increased awareness that in natural environments pure cultures are the exception rather than the rule. If we are to achieve a meaningful understanding of how microbial ecosystems function *in situ*, greater emphasis must be placed on the study of microbial communities.

The complex structural and dynamic nature of microbial ecosystems makes *in situ* studies a daunting task, and this, coupled with constantly changing environmental conditions during the period of investigation, makes valid interpretation of field data a major problem. Until recently the experimental methodologies available to study microbial ecosystems *in situ* have been insufficiently sensitive or discrete to probe these habitats without destroying their structural integrity. New methods for determining bacterial biomass have been developed which do not require the quantitative removal of cells for enumeration. These methods depend upon the measurement of specific biochemical components which are restricted to particular sub-groups of the microbial community under investigation. Muramic acid (Moriarty 1977), lipopolysaccharide (Saddler & Wardlaw 1980; Parker *et al.* 1982), phospholipids (King *et al.* 1977; White *et al.* 1979) and teichoic acids (Gehron *et al.*

0-632-01793-7

Industrial
Microbiological Testing

1982) have been identified as specific biochemical 'signature' components which can be measured with a high degree of precision (sensitivity 10^{-11}–10^{-12} moles). Community composition can similarly be determined by extension of the 'signature' lipid techniques to differentiate prokaryote cells from eukaryotes and aerobes from anaerobes (White 1983). The metabolic activity of microbial communities can be determined by measuring the rate of incorporation of radiolabelled substrates, e.g. [^{14}C]glucose, [^{14}C]glutamate, [^{3}H]thymidine, etc. (Sorokin & Kadota 1972; Dawson & Gocke 1978; Fuhrman & Azam, 1980). Needle microelectrodes have been successfully used to measure physico-chemical gradients (pH, Eh, S^{2-}, O_2) in benthic communities (Jorgensen *et al.* 1979; Jorgensen 1983). Whilst these new techniques have greatly increased our understanding of how microbial communities function *in situ* it is frequently necessary to develop simple laboratory model systems to elucidate the key processes and interactions involved. The principal advantage of laboratory based systems is that environmental variables such as O_2, pH, nutrient concentrations and growth can be precisely regulated for the duration of the experiment. As a consequence of this degree of control, individual parameters can be varied in an ordered and predictable manner, thereby facilitating interpretation of the data. Laboratory model systems, however, must be relevant to the ecosystem under investigation if they are to advance our knowledge of how microbial communities function *in situ*. In this paper we describe the application of biphasic gel stabilized models to the study of microbial degradation of North Sea oil under anaerobic conditions.

Microbiological Problems Associated with the Petroleum Industry

Micro-organisms cause major operational and economic problems in the oil industry. These problems include the plugging of oil-bearing strata with bacteria, the blocking of oil-pipelines, the contamination of water used in secondary oil recovery and contamination of oil storage systems. Additionally, problems may arise in connection with the corrosion of equipment, the production of highly toxic gases such as H_2S and the fouling of oil rig structures (drilling legs, etc.). The economic cost of these activities of micro-organisms is difficult to estimate precisely, but figures for the cost of maintaining production platforms and their associated equipment are high (Skrebowski 1978). Hamilton (1983) quotes figures of between £300M and £500M per annum for North Sea operations and considers that the total maintenance costs over the lifespan of an oil-field may be in the order of 50% of the initial capital costs. Hydrocarbons are the primary source of nutrients for micro-organisms in these habitats, and there is considerable interest in how these carbon sources are metabolized, particularly under anaerobic conditions. The principal group of anaerobics of interest to the petroleum industry are sulphate-reducing bacteria, since they have been implicated in the corrosion of drill structures and pipelines, the production of toxic H_2S gas and souring of stored crude-oil. Until recently, sulphate-reducing bacteria were considered to be a nutritionally restricted group of strictly anaerobic, heterotrophic micro-organisms belonging to two genera, *Desulfovibrio* and *Desulfotomaculum* (Postgate 1984). However, Widdel (1980) has described several new genera of sulphate-reducing bacteria which are metabolically more versatile, with individual species able to grow on C_2–C_{18} fatty acids and aromatic compounds such as benzoate and phenol. The characteristic feature of all sulphate-reducing bacteria is their ability to carry out anaerobic respiration using SO_4^{2-} as the terminal e^- acceptor. Whilst sulphate-reducing bacteria are commonly isolated from crude oil and secondary oil recovery injection waters, there is no unequivocal experimental evidence to show that these micro-organisms can use hydrocarbons as carbon and energy sources (Hardy 1981; Hamilton 1983; Postgate 1984). It is more probable that sulphate-reducing bacteria form one component of complex microbial communities, some of whose members are able to utilize hydrocarbons as primary carbon and energy sources. The fermentation end-products produced by these hydrocarbon utilizers are in turn further oxidized by other members of the microbial community. When SO_4^{2-} is present, as for example in the North Sea, then sulphate-reducing bacteria are primarily responsible for these terminal oxidation steps with the concomitant production of H_2S. To determine the nature of these interactions between different physiological groups of micro-organisms we have modified the biphasic gel model system first described by Wimpenny *et al.* (1981).

Application of Gel-stabilized Model Systems

Adding gelling agents such as agar to liquid media prevents solute movement by convection or mechanical means, leaving molecular diffusion processes as the sole mechanism for solute transfer. This simple concept has been successfully exploited by Wimpenny and co-workers (Wimpenny & Whittaker 1979; Wimpenny *et al.* 1981, 1983) to study microbial growth in solute gradients. For example, if the agar source layer contains glucose and the gels are incubated aerobically, then the microbial populations migrate and develop as discrete bands in the opposing glucose and O_2 gradients, which introduces an element of spatial heterogeneity into the gel system. The spatial distribution of individual microbial populations, obtained from either a natural or defined mixed culture inoculum, is dependent upon individual cells migrating to horizons within the gel where physico-chemical conditions are optimal for growth. By employing needle electrode systems dissolved, O_2, Eh, pH and S^{2-} concentrations can be determined. Microbial growth can be determined either using a gel scanner or by viable count determinations on gel cores which have been aseptically removed, sliced into thin sections and homogenized. Wimpenny *et al.* (1981) have used gel models successfully to investigate the ecology of oil storage water bases where corrosion problems are due to the activities of sulphate-reducing bacteria.

Construction of Gel-stabilized Model Systems to Study Oil Degradation under Anaerobic Conditions

Anaerobic biphasic gel systems were established in 250 ml heat-sterilized beakers. The semi-solid base layer had the following composition ($g.l^{-1}$); Na_2HPO_4, 0.5; $MgSO_4.7H_2O$, 0.1; NaCl, 11.6; nitrilotriacetic acid, 0.05; NH_4Cl, 1.0; KCl, 0.5; yeast extract, 0.1; trace elements, 1 ml of solutions A, B and C (Brown & Stanley 1972). Washed purified agar (Oxoid Ltd) was added to give a final concentration of 0.4% w/v. The pH of the medium was adjusted to 7.2 before autoclaving at 121°C for 15 min. After cooling to 45°C, 125 ml volumes of the medium supplemented with 0.5 ml volumes of an appropriate inoculum (oil-polluted sediment or defined mixed bacterial populations) were poured into sterile beakers and allowed to gel under an atmosphere of oxygen-free nitrogen to maintain anaerobic conditions. North Sea oil (B.P. Forties Field) was then carefully poured on top of the semi-solid agar layer to complete the experimental system. The beakers were covered with aluminium foil to maintain sterility and reduce the loss of volatile oil fractions. The gels were incubated at 23°C and examined at regular time intervals for the development of bands of microbial growth.

Preparation of Bacterial Inoculum

Defined mixed populations of bacteria were used to inoculate the semi-solid agar base layer. The bacterial isolates included a saccharolytic *Clostridium* spp., designated *Cl. butyricum* SS6 (Keith *et al.* 1982a), a sulphate-reducing bacterium belonging to the genus *Desulfovibrio* and designated *Dv. desulfuricans* DT 101 (Keith *et al.*, 1982b), a hydrocarbon degrading *Vibrio* spp., designated HU-5, isolated from oil-polluted sediment in gel enrichments and *E. coli* K12. *Cl. butyricum* SS6 was grown in a modification of Skinner's medium (Keith *et al.* 1982a), *Dv. desulfuricans* DT101 in Postgate's medium E (Postgate 1984), *Vibrio* HU-5 and *E. coli* K12 in Tryptone Soy broth. With the exception of *E. coli* K12, which was incubated at 37°C, the remaining isolates were grown at 23°C for 24 h. The paired combinations of isolates included: *Cl. butyricum* SS6/*Dv. desulfuricans* DT101; *Vibrio* HU–5/*Dv. desulfuricans* DT101 and *E. coli* K12/*Dv. desulfuricans* DT101.

Microbiological and Chemical Analysis of Gel Systems

Sampling the gels. A sterile 10 mm internal diameter cork borer was used to remove samples from each gel. The gel core was carefully extruded, under aseptic conditions, from the lumen of the cork borer and sliced into 2.5 mm thick sections. Viable count determinations involved the release of cells by homogenization in 9 ml volumes of 0.9% w/v physiological saline.

Viable count determinations. Aerobic bacterial counts from each gel horizon were prepared as serial ten-fold dilutions and 0.1 ml volumes from each dilution plated out on Tryptone Soy Agar. Anaerobic counts (of *Cl. butyricum* SS6 and *Dv. desulfuricans* DT101) were performed using the agar shake technique developed by van Niel (1931). The sulphate-reducing bacterial populations were enumerated using growth medium E (Postgate 1984) and populations of saccharolytic clostridia using Skinner's medium as modified by Keith *et al.*, (1982a).

Analytical methods using needle electrodes. Needle electrodes for measuring pH and Eh were mounted on a micromanipulator (W. Prior & Co. Ltd, Bishop's Stortford, Herts.) and lowered into the semi-solid agar layer at measured intervals. The pH in the semi-solid agar layer was measured using a needle electrode (MI-405; Microelectrodes Inc., Londonderry, N.H., U.S.A.) with a calomel reference electrode (Russell pH, Auchtermuchty, Fife, Scotland) connected by a KCl salt bridge to an EIL model 7055 pH meter (Kent Instruments Ltd, Chertsey, Surrey). Redox potential (Eh) measurements were made using a bare platinum electrode and a calomel reference electrode connected via a KCl salt bridge to the EIL pH meter. Free sulphide was measured colorimetrically according to the method of Truper & Schlegel (1964).

Performance of Gel Model Systems

In initial experiments the gels were used as enrichment systems for isolating hydrocarbon-utilizing bacteria. The inoculum was oil-polluted sediment (0.5 ml) from the Tay Estuary. Data presented in Tables 1 and 2 show the diversity of bacterial types isolated using these procedures and the spectrum of pure hydrocarbons they were able to utilize. The majority of the isolates obtained (84%) were Gram-negative rod-shaped bacteria which were identified to genus level using a limited number of morphological and biochemical tests. Facultative anaerobes identified as *Vibrio* spp. were the most frequently isolated hydrocarbon utilizers (45%) and metabolically the most versatile, being able to utilize a range of straight chain paraffins as well as aromatics such as benzene (Table 2). Isolates HU-19 and HU-20 which were identified as *Klebsiella* spp. were unusual in that they were unable to utilize any of the hydrocarbons tested, suggesting that they were dependent on metabolic end products produced by the primary hydrocarbon utilizers for growth. A noticeable feature of these gel enrichments was the appearance of bands of microbial growth in the semi-solid agar layer immediately underlying the oil source layer (Fig. 1). These bands of growth are similar to those reported by Wimpenny *et al.* (1981) who noted that in their experimental system band formation

TABLE 1. *Identification of hydrocarbon-utilizing bacteria isolated from oil polluted sediments in the Tay estuary using biphasic gel enrichments*

Isolate	Morphology	Motility	Gram reaction	Oxidase	Catalase	H & L O/F test	Genus
HU1	R	+	−	+	+	O	*Pseudomonas*
HU2	R	+	−	+	+	O	*Pseudomonas*
HU3	R	+	−	+	+	O	*Pseudomonas*
HU4	R	−	−	+	+	−	*Acinetobacter*
HU5	R	+	−	+	+	F	*Vibrio*
HU6	R	+	−	+	+	F	*Vibrio*
HU7	R	+	−	+	+	−	*Acinetobacter*
HU8	R	+	+	+	+	F	*Bacillus*
HU9	R	+	+	±	+	F	*Bacillus*
HU10	R	+	+	+	+	F	*Bacillus*
HU11	R	+	−	+	+	F	*Vibrio*
HU12	R	+	−	+	+	F	*Vibrio*
HU13	R	+	−	+	+	F	*Vibrio*
HU14	R	+	−	+	+	F	*Vibrio*
HU15	R	+	−	+	+	F	*Vibrio*
HU16	R	+	−	+	+	F	*Vibrio*
HU17	R	+	−	+	+	F	*Vibrio*
HU18	R	+	−	+	+	−	*Alcaligenes*
HU19	R	−	−	−	+	F	*Klebsiella*
HU20	R	−	−	−	+	F	*Klebsiella*

R = rod shaped.

was only associated with bacteria capable of anaerobic growth. After 6 weeks incubation, the banding in the gels became progressively more diffuse and gradually merged until individual bands could no longer be distinguished. At the interface between the semi-solid agar layer and the oil source layer the film of bacterial growth became so extensive that it extended into the oil and formed an emulsion. Under these conditions discrete oil droplets were formed (Fig. 2). These discrete oil droplets increased the susceptibility of the oil to degradation, since a larger surface area was exposed to the hydrocarbon-oxidizing bacteria (Floodgate, 1978).

Application of Biphasic Gel Systems to the Study of Oil Degradation by Defined Microbial Populations

In initial experiments a series of gels was set up with single pure cultures of *Cl. butyricum* SS6, *Vibrio* HU-5, *Dv. desulfuricans* DT101 and *E. coli* K12. Good growth was observed with *Cl. butyricum* SS6 and *Vibrio* HU-5 but not with *Dv. desulfuricans*

TABLE 2. *Pattern of hydrocarbon utilization by hydrocarbon-oxidizing bacteria isolated from oil-polluted Tay estuary sediments by gel enrichments. Each isolate was grown in the basal mineral salts medium supplemented with 1% v/v of each individual hydrocarbon and incubated at 23°C for up to 10 days. Growth was estimated by comparison with inoculated controls containing no carbon source.*

	Hydrocarbon source					
Isolate	Benzene	n-Octane	n-Decane	n-Dodecane	n-Tetradecane	n-Hexadecane
HU1	+/−	+/−	+/−	+/−	+	+
HU2	+/−	+/−	+	+	+	+
HU3	+	++	++	++	++	++
HU4	−	+/−	++	++	++	++
HU5	+	+	+	++	++	++
HU6	+	−	++	++	++	++
HU7	−	+	+/−	+/−	+/−	+
HU8	+/−	+	++	+	++	++
HU9	+	+	++	++	++	++
HU10	+/−	+/−	+	++	++	+
HU11	+/−	+/−	+/−	+/−	+	+
HU12	+	+	++	++	++	+
HU13	+	+	++	++	++	++
HU14	+	+	+	+/−	+/−	+
HU15	+	+	++	++	+	+
HU16	+	+	+/−	+	++	++
HU17	+	+/−	+/−	+	+	+
HU18	−	−	+/−	+/−	+	+
HU19	−	−	−	−	−	−
HU20	−	−	−	−	−	−

++, strong growth; +, good growth; +/−, slight growth; −, no growth.

DT101 or *E. coli* K12. Fig. 3 shows the pattern of growth which developed when *Cl. butyricum* SS6 was grown under these conditions. When defined mixed populations were used as the inoculum, good growth was observed with *Cl. butyricum* SS6/*Dv. desulfuricans* DT101 (Fig. 4) and *Vibrio* HU-5/*Dv. desulfuricans* but not for the combination of *E. coli* K12/*Dv. desulfuricans* DT101. The pattern of growth which developed with defined mixed cultures was similar irrespective of the combinations used. The hydrocarbon utilizer (*Cl. butyricum* SS6 or *Vibrio* HU-5) always grew at the interface between the oil layer and the semi-solid layer and once this population had become established (typically 3–4 days at 23°C) growth of *Dv. desulfuricans* DT101 occurred forming a black band immediately below the primary hydrocarbon utilizer. This spatial arrangement of the two populations for the defined mixed culture of *Cl. butyricum* SS6/*Dv. desulfuricans* DT101 is shown more clearly in Fig. 5. Maximum populations of the hydrocarbon utilizer were recorded at 0–5 mm depth, whilst highest numbers of the sulphate-reducing bacteria were observed in a broad band extending from 15 to 25 mm below the agar surface. Similar population

FIG. 1. Development of multiple bands of microbial growth in biphasic gel enrichments incubated at 23°C for 10 days. Inoculum was 0.5 ml of oil-polluted estuaries sediment.

distributions were recorded with the defined mixed cultures of *Vibrio* HU-5/*Dv. desulfuricans* DT101 (data not presented). Profiles of physico-chemical gradients in these gels show that major changes occurred in Eh, pH and soluble S^{2-} concentrations with increasing depth. The Eh of the semi-solid agar layer decreased from −85 mV at the surface to −220 mV at 15 mm depth and then showed little further change with depth. Similarly, marked changes in pH values occurred with depth. At the surface of the semi-solid agar layer (0–5 mm depth) the pH was 5.25 due to the production of acid fermentation products by *Cl. butyricum* SS6 but the medium became progressively more alkaline with increasing depth (pH 7.43) due to the generation of free S^{2-} by *Dv. desulfuricans* DT101. Maximum free S^{2-}

FIG. 2. Formation of discrete oil droplets at the oil layer/semi-solid agar layer interface due to the production of emulsifying agents by the primary hydrocarbon-utilizing bacteria. Gel system incubated at 23°C for 15 days.

(a) (b)

FIG. 3. Growth of *Cl. butyricum* SS6 (a) at the oil layer/semi-solid agar layer interface compared with an uninoculated gel (b). Gel system incubated at 23°C for 3 days.

FIG. 4. Growth of a defined mixed culture of *Cl. butyricum* SS6/*Desulfovibrio desulfuricans* DT101 in the biphasic gel system. The white band immediately underlying the oil layer is the clostridium and the black zone the sulphate-reducing bacterium. Gel system incubated at 23°C for 7 days.

concentrations were recorded between 15 and 20 mm depth and were coincident with maximum cell densities observed for *Dv. desulfuricans* DT101 (Fig. 5). These data show that an obligate commensal relationship developed between the primary hydrocarbon utilizer (*Cl. butyricum* SS6 or *Vibrio* HU-5) and the sulphate-reducing bacterium. No inhibition was observed of either the *Cl. butyricum* SS6 or *Vibrio* HU-5 populations by free S^{2-} produced by the sulphate-reducing bacterium. Indeed, batch culture experiments have demonstrated that *Cl. butyricum* SS6 will tolerate up to 10 mm S^{2-} without any significant effect on growth rate.

Discussion

Biphasic gel stabilized model systems, as described in this chapter, represent a simple yet powerful technique for investigating the spatial organization of microbial ecosystems. Model systems such as these combine spatial heterogeneity with stability

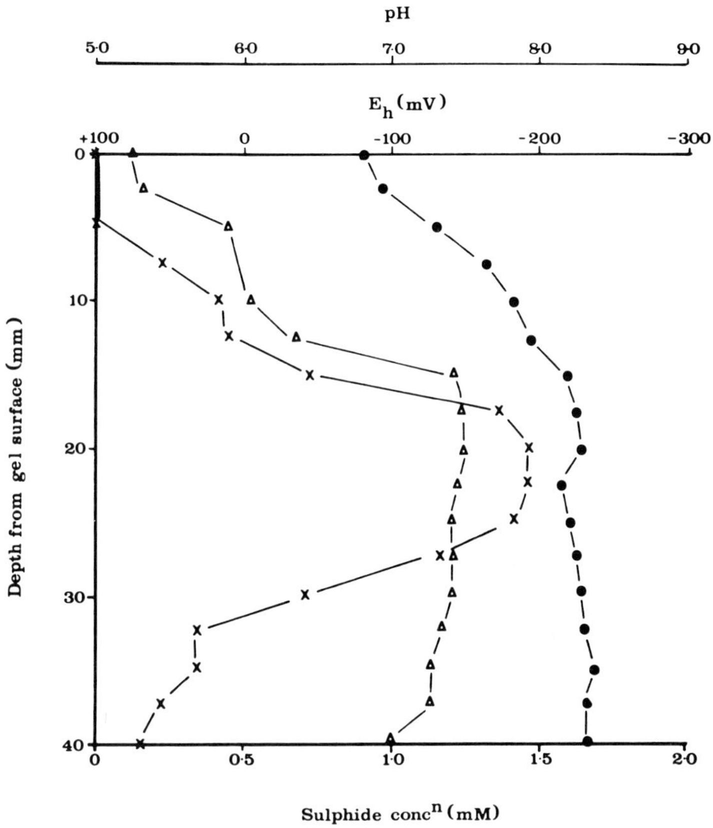

FIG. 5. Physio-chemical gradient profiles which developed in gels inoculated with defined mixed populations of *Cl. butyricum* SS6/*Dv. desulfuricans* DT101. Gels incubated at 23°C for 7 days. pH (△), E_h (●), soluble suphide concentration (X).

and reproducibility. By using microelectrodes the physico-chemical gradients which develop can be accurately measured with the minimum of mechanical disturbance, thereby maintaining the spatial heterogeneity of the system.

The oil/agar gel system employed in this investigation model closely resembles environmental conditions found in oil storage tanks and in oil wells in the North Sea. Wimpenny *et al.* (1981) successfully used biphasic gel systems to study the physico-chemical gradients in oil storage tanks contaminated with water. In this investigation we have extended this experimental approach to study the interactions between different groups of micro-organisms and how these may influence the degradation of North Sea oil under anaerobic conditions. Wilkinson (1982) reported that the rate at which sulphate-reducing bacteria were able to grow in the stagnant

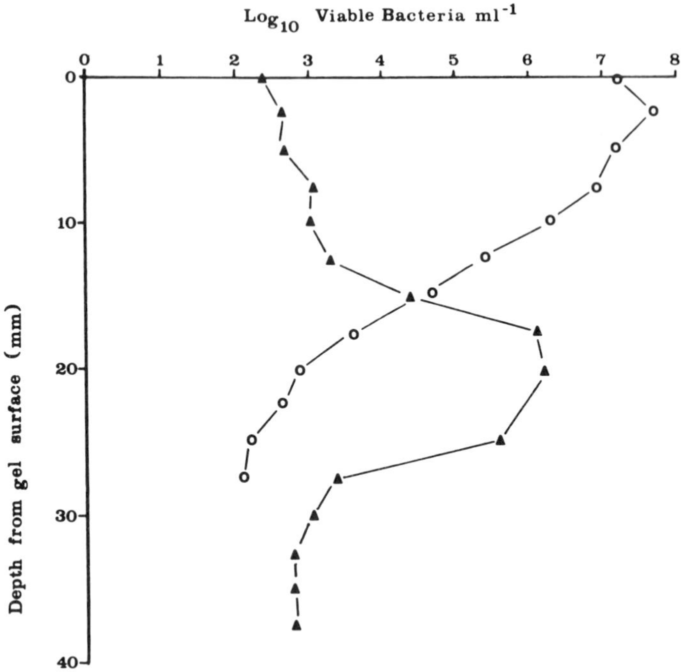

FIG. 6. Population profiles of *Cl. butyricum* SS6/*Dv. desulfuricans* DT101 which developed in the biphasic gel systems when inoculated with defined mixed populations of both bacteria. Gels incubated at 23°C for 7 days. *Cl. butyricum* SS6 (○); *Dv. desulfuricans* DT101 (▲).

water of oil storage cells in offshore structures was dependent upon the presence of hydrocarbons. When no hydrocarbons were available the mean generation time for the sulphate-reducing bacteria was in the order of 144–192 h and free S^{2-} concentrations were in the order of 5 mg l^{-1}. In contrast, when hydrocarbons were present the mean generation time was 10–15 h and free S^{2-} levels were between 50 and 100 mg l^{-1}. The question which must therefore be addressed is whether or not sulphate-reducing bacteria are capable of oxidizing hydrocarbons. At the present time the general consensus is that sulphate-reducing bacteria do not utilize hydrocarbons (Wilkinson 1982; Hamilton 1983; Postgate 1984). In this study we have equally failed to demonstrate that sulphate-reducing bacteria will use hydrocarbons directly. However, we have unequivocally demonstrated, using defined mixed populations of hydrocarbon utilizing bacteria (*Cl. butyricum* SS6 and *Vibrio* HU-5) and *Dv. desulfuricans* DT101, that good growth of the sulphate-reducing bacterium occurred under these conditions. The primary hydrocarbon utilizers always grew at the oil/semi-solid agar layer interface in the biphasic gel model systems, and there is good evidence to show that they produced emulsifying agents which significantly increased the surface area of contact between the bacterial cells and the oil droplets which were produced.

In summary, the examples of biphasic gel systems described here illustrate the potential application of these simple laboratory models to evaluate the fundamental principles governing the function of complex microbial ecosystems. These include delineating habitat requirements for individual physiological types of micro-organisms, as enrichment and isolation systems and for the investigation of interactions between different physiological groups of micro-organisms. Whilst gel stabilized laboratory model systems offer considerable potential as a research tool for investigating microbial interactions, they nonetheless do suffer from a number of problems and limitations. For example, they are closed systems with respect to solutes, they lack particulate matter and are not subject to bioperturbation. Despite these obvious disadvantages they nevertheless permit the analysis of microbial processes and interactions which are presently unobtainable by other means.

References

Brown, C. M. & Stanley, S. O. 1972 Environment mediated changes in the cellular content of the 'pool' constituents and their associated changes in cell physiology. *Journal of Applied Chemistry and Biotechnology* **29**, 447–454.

Dawson, R. & Gocke, K. 1978 Heterotrophic activity in comparison to the free amino acid concentrations in Baltic Sea water samples. *Oceanology Acta* **1**, 45–54.

Floodgate, G. D. 1978 The formation of oil emulsifying agents in hydrocarbonclastic bacteria. *Microbial Ecology*, eds Loutit, M. W. & Miles, J. A. R., pp. 573–575. Berlin: Springer-Verlag.

Fuhmran, J. A. & Azam, F. 1980 Bakterioplankton secondary production estimates for coastal waters of British Columbia, Antarctica and California. *Applied and Environmental Microbiology* **39**, 1085–1095.

Gehron, M. J., Moriarty, D. J. W., Smith, G. A. & White, D. C. 1982 Determination of the Gram negative, Gram positive bacterial content of sediments. *Applied and Environmental Microbiology* **44**, 1041–1049.

Hamilton, A. 1983 Sulphate reducing bacteria and the offshore oil industry. *Trends in Biotechnology* **1**, 36–40.

Hardy, J. A. 1981 The enumeration, isolation and characterisation of sulphate reducing bacteria from North Sea waters. *Journal of Applied Bacteriology* **51**, 505–516.

Jorgensen, B. B. 1983 Ecology of the bacteria of the sulphur cycle with special reference of anoxic-oxic interface environments. *Philosophical Transactions of the Royal Society*, B. **298**, 543–561.

Jorgensen, B. B., Kuenen, J. G. & Cohen, Y. 1979 Microbial transformations of sulphur compounds in a stratified lake (Solar Lake), Sinai. *Limnology and Oceanography* **24**, 799–822.

Keith, S. M., Herbert, R. A. & Harfoot, C. G. 1982b. Isolation of new types of sulphate reducing bacteria from estuarine sediments. *Journal of Applied Bacteriology* **53**, 29–33.

Keith, S. M., Macfarlane, G. T. & Herbert, R. A. 1982a. Dissimilatory nitrate reduction by a strain of *Clostridium butyricum* isolated from estuarine sediments. *Archives of Microbiology* **132**, 62–66.

King, J. D., White, D. C. & Taylor, C. W. 1977 Use of lipid composition and metabolism to examine structure and activity of estuarine detrital microflora. *Applied and Environmental Microbiology* **33**, 1177–1183.

Moriarty, D. J. W. 1977 Improved method using muramic acid to estimate biomass of bacteria in sediments. *Oecologica* **26**, 317–323.

Parker, J. H., Smith, G. A., Fredrickson, H. L., Vestal, J. R. & White, D. C. 1982 A sensitive assay for Gram negative bacteria in sediments based on hydroxy-fatty acids from lipopolysaccharide lipid A. *Applied and Environmental Microbiology* **44**, 1170–1177.

Postgate, J. R. 1984 *The Sulphate Reducing Bacteria.* London: Cambridge University Press.

Saddler, J. N. & Wardlaw, A. C. 1980 Extraction, distribution and biodegradation of bacterial lipopolysaccharides in estuarine sediments. *Antonie van Leeuwenhoek Journal of Microbiology* **46**, 27–39.

Skrebowski, C. 1978 Is Britain's oil industry on shaky legs? *New Scientist* March, 714–716.

Sorokin, Y. I. & Kadota, H. 1972 *Microbial Production and Decomposition in Fresh Waters, IBP Handbook* No. 23. Oxford: Blackwell Scientific Publications.

Truper, H. G. & Schlegel, H. G. 1964 Sulfur metabolism in Thiorodaceae. 1. Quantitative measurements on growing cells of *Chromatium okenii. Antonie van Leeuwenhoek* **30**, 225–238.

van Niel, C. B. 1931 On the morphology and physiology of the purple and green sulphur bacteria. *Archiv für Mikrobiologie* **3**, 1–112.

White, D. C. 1983 Analysis of microorganisms in terms of quantity and activity in natural environments. *Microbes in their Natural Environments: Symposia of the Society for General Microbiology*, **34**, eds Slater, J. H., Whittenbury, R. & Wimpenny, J. W. T., pp. 37–66. London: Cambridge University Press.

White, D. C., Davis, W. M., Nickels, J. S., King, J. D. & Bobbie, R. J. 1979 Determination of the sedimentary microbial biomass by extractable lipid phosphate. *Oecologica* **40**, 51–62.

Widdel, F. 1980 Anaerober Abbau von Fettsauren und Benzosaure durch neu isoliert Arten Sulfat-reduzierender Bakterien. Doctoral Thesis, University of Gottingen, West Germany.

Wilkinson, T. 1982 An environmental programme for offshore oil operations. *Chemistry and Industry* **1**, 115–123.

Wimpenny, J. W. T., Coombs, J. P., Lovitt, R. W. & Whittaker, S. 1981 A gel stabilised model ecosystem for investigating microbial growth in spatially ordered solute gradients. *Journal of General Microbiology* **127**, 277–287.

Wimpenny, J. W. T., Lovitt, R. W. & Coombs, J. P. 1983 Laboratory model systems for the investigation of spatially and temporally organised microbial ecosystems. *Microbes in their Natural Environments: Symposia of the Society for General Microbiology* **34**, eds Slater, J. H., Whittenbury, R. & Wimpenny, J. W. T., pp. 66–117. London: Cambridge University Press.

Wimpenny, J. W. T. & Whittaker, S. 1979 Microbial growth in gel stabilised gel gradients. *Society for General Microbiology Quarterly* **6**, 80.

A Method for Testing Light Distillate Fuels

B. N. Herbert,[1] E. C. Hill,[2] P. D. Oates,[3] D. Powell,[4]
Jean L. Shennan[5] and R. Whittle[6]
[1]*Shell Research Ltd, Sittingbourne;* [2]*E. C. Hill & Associates, Cardiff;* [3]*Castrol Research Laboratories, Pangbourne;* [4]*Admiralty Research Establishment Dockyard Laboratory, Portsmouth;* [5]*BP Research Centre, Sunbury-on-Thames;* [6]*Vosper-Thornycroft (UK) Ltd, Southampton.*

Any petroleum fuel storage tank which also contains a layer of water underneath the fuel, a 'water bottom', will harbour micro-organisms capable of proliferation, if physical conditions allow, into dense mats of microbial biomass floating at the oil/water interface. Many species of filamentous fungi, yeasts and bacteria are capable of assimilating hydrocarbons, especially straight chain alkanes and lower molecular weight cyclic and aromatic molecules found in crude and refined oil products. Apart from their carbon content, refined fuels are relatively poor nutritionally, but the other microbial requirements for growth on hydrocarbons—dissolved oxygen, an available nitrogen source, mineral salts and trace elements—are invariably present in fuel tanks as the turn-over of the fuel phase frequently renews the supply of these factors to the micro-organisms in the relatively static water phase.

Microbial interfacial growth, if allowed to develop, will entrap dust, grit, metal oxides and metal swarf in addition to microbial biomass. Microbial activity is also likely to produce polymeric substances, increasing the bulk of the interfacial mass, and surface active metabolites which may create a stable emulsified water-in-oil phase between the bulk fuel and the zone of microbial activity. The floating biomass will eventually fall to the base of the storage tank to accumulate as sludge together with organic and inorganic debris. Anaerobic conditions under this sludge blanket and under any wall growth of microbial slime will provide a favourable environment for the activity of sulphate-reducing bacteria.

The detrimental consequences of uncontrolled microbial growth in fuels have been reviewed by several authors, including Davis (1967), Sheridan *et al.* (1972), Park (1975) and Hill (1984). Under most circumstances, contact time between fuel and an underlying water phase will be too short for any chemical change in the fuel itself associated with microbial growth. However, the potential for microbial contamination of the fuel phase is always present.

Industrial Microbiological Testing
0–632–01793–7

In the course of distribution from refinery production unit to fuel burner unit, fuel oils will pass through a number of storage tanks of various sizes for differing residence times. At each stage the presence of water, as condensation on tank walls or as water bottoms formed by accumulated condensate or rain water leakage, is inevitable unless stringent housekeeping practices are carried out. Viable cells or fungal spores picked up by the fuel phase will be carried through each subsequent stage of storage along the distribution line to contaminate water bottoms throughout the system. Only in the case of jet aircraft fuel (ATK) do the safety hazards of fuel contamination demand the high standards of tank maintenance, fuel filtration and dewatering on transfer together with fuel quality monitoring, necessary to keep particulate matter and viable microbial cells or spores at minimal levels.

In other fuels, the economically serious aspects of microbial contamination in storage tanks are associated with the carriage of interfacial material into the distribution system with the fuel off-take resulting in clogged filters and coalescers and blocked pipes and valves. Other effects include the emulsification of water in the fuel phase affecting its burn or ignition properties and microbially mediated corrosion of tank walls by sulphate-reducing activity underneath sludge and slime layers.

The control of micro-organisms in stored fuels is, first and foremost, by careful management of the tankage combined with good design of the tanks and connecting pipework systems, to prevent ingress of rain water and allow frequent and complete drainage of condensate or any water admixed with the fuel charge. Biocide treatment of water bottoms is possible but is seldom done routinely. Biocides have to be selected against many criteria of suitability, including cost, compatibility with the burn characteristics of the fuel and the human and environmental safety of the discharged treated water (Rogers & Kaplan 1968; Hill 1976; Klemme & Neihof 1976; Smith & Crook 1982).

A situation over which little or no microbial control can be exercised is the carriage and storage of fuel in sea-water ballasted tanks on ships. Problems associated with light diesel fuel (DIESO) contamination aboard gas turbine warships have been reported by U.K. and U.S. Navies (Houghton & Gage 1979; Neihof & May 1983). Interfacial microbial growth in shipboard fuel tanks is not a new problem but has been highlighted in recent years by tighter specifications required for fuel burnt in turbines. This has led to the fitting of fine filtration and water removal equipment in vessels with gas turbine propulsion which are consequently more susceptible to blockage. Oil-rig supply vessels and storage tanks in oil production platforms are also sea-water ballasted and similar problems of blockage and clogging can arise in diesel generators on the platforms if supplied with such contaminated fuel.

In the course of routine testing of fuels, analysis of contamination problems or laboratory screening of biocides for fuels, a method for estimating microbial contamination in the fuel phase is required. Most methods are based on the membrane filtration method of Hazzard (1967) for viable counts, but microscopy and direct culturing of samples of heavily contaminated fuel will yield useful qualitative data (Neihof & May 1982; Rogers & Kaplan 1982). Where rapid analysis of numerous samples is required, membrane filtration followed by direct counting of hyphal fragments can be used. Test kits for hydrocarbon fuel systems have also been used to monitor contamination in the field (Bailey & May 1979).

Standard methods for microbial analysis of hydrocarbon/water systems are virtually impossible to draw up, as there are many variables which may make handling of the sample difficult. Following a series of interlaboratory tests carried out on reference fuels, the authors have combined their experience on fuel testing into a method which is described and discussed in this paper. It is presented as the best approach to light fuel testing as agreed by the collaborating laboratories.

The method is suitable for light distillate fuels, including: aviation turbine kerosine (ATK, AVTUR, Jet-Al); aviation turbine gasoline (AVTAG, JP.4); aviation gasoline (AVGAS); high-speed diesel engine fuels (DIESO, DERV, Gas-oil); central heating oils (Gas-oil).

The Method

Sampling

A representative sample, preferably 400–500 ml, should be taken into a sterile container. If plastic containers are used, they must be fuel-resistant and pre-sterilized. If metal-capped bottles or cans are used, the cap-liner must be of fuel-resistant material.

Samples should be kept cool and transported to the laboratory as quickly as possible. The number of days in transit should be noted.

Water-phase analysis

If the sample contains free water it should be allowed to settle and separated aseptically from the fuel phase. The water phase can then be examined by microscopy and conventional culture techniques.

Fuel-phase analysis for viable fungal spores and fragments

Filtration. Filter a range of volumes through pre-sterilized 47 mm diameter membrane filters of 0.45 μm pore size held in pre-sterilized membrane funnels.

Suitable fuel volumes for filtration would be aliquots of 1, 10 and 100 ml. Choice of volumes will also be dictated by the level of contamination expected and the filterability of the fuel.

Duplicate tests should be carried out on each selected volume subsample.

Detergent wash. Wash the fuel sample through the membrane with a dispersant solution. Depending on filterability, the membranes should be washed with as many units of 10 ml dispersant as practicable, up to a recommended total of 100 ml.

Suitable dispersants include BP 1100X (British Petroleum), Corexit 7664 (Esso Chemicals) or Dispersant LTX (Shell Chemicals), prepared as 0.1% v/v aqueous solutions and sterilized by autoclaving.

Saline rinse. To remove dispersant from the membrane, follow with a wash of sterile physiological saline (0.85% v/v solution of sodium chloride in water). Again, depending on filterability, the membranes should be rinsed with as many units of 10 ml saline as practicable, up to a total of 30 ml.

Incubation of membranes

Transfer the membranes aseptically face up to petri dishes containing *c.* 15 ml agar medium.

Two agar media are used for fungal counts: Malt Extract Agar (e.g. Oxoid CM 59) and Sabouraud Dextrose Agar (e.g. Oxoid CM 41). An antibiotic, chlortetracycline or cyclohexamide (0.001% w/v), is incorporated in these media to suppress bacterial growth.

Incubate at 25 ± 2°C for 4–7 days. Examine daily and count colony forming units and identify the colonies as far as is practicable.

Discussion

Practical experience has shown that standard procedures cannot be imposed on the testing of light distillate fuels. Filterability—how much of the sample can be passed through the membrane before blockage occurs—is a particular problem and depends on the composition of the fuel and the chemical content of the associated water. The batchwise addition of both the dispersant wash and the saline rinse also recognizes that the operator may not be able to get the full recommended volume through the membrane.

It is important to separate the fuel and water phase in a water-contaminated sample, and make careful interpretation of the results of corresponding analyses of the water and fuel phases. Many bacteria do not survive when dispersed in fuel. A sample containing free water may therefore appear to have different microbial populations in the oil and water if they are tested separately.

Both mixed cellulose ester filters and nuclepore filters can be used according to individual preference.

The procedure as described is designed to enumerate viable spores and hyphal fragments of fungi and viable yeast cells. If bacteria are to be recovered from fuels, a 0.22 μm porosity membrane and a suitable nutrient agar, e.g. Oxoid Plate Count Agar CM 325, should be used.

References

Bailey, C. A. & May, M. E. 1979 Evaluation of microbiological test kits for hydrocarbon fuel systems. *Applied and Environmental Microbiology* **37**, 871–877.

DAVIS, J. B. 1967 *Petroleum Microbiology*. New York: Elsevier.

HAZZARD, G. F. 1967 The detection of microorganisms in petroleum products. *Journal of the Institute of Petroleum* **53**, 267–274.

HILL, E. C. 1976 Evaluation of biocides for use with petroleum products. *Process Biochemistry* **11**, 36–38, 48.

HILL, E. C. 1984 Biodegradation of petroleum products. In *Petroleum Microbiology*, ed. Atlas, R. M., pp. 579–617. New York: Macmillan.

HOUGHTON, D. R. & GAGE, S. A. 1979 Biology in ships. *Transactions of the Institute of Marine Engineers* **91**, 1–7.

KLEMME, D. E. & NEIHOF, R. A. 1976 An evaluation in large-scale test systems of biocides for control of sulphate-reducing bacteria in shipboard fuel tanks. Naval Research Laboratory Memorandum Report No. 3212, Washington, D.C.

NEIHOF, R. A. & MAY, M. E. 1982 Microbial contamination of naval fuels. *Developments in Industrial Microbiology* **23**, 167–176.

NEIHOF, R. A. & MAY, M. E. 1983 Microbial and particulate contamination in fuel tanks on naval ships. *International Biodeterioration Bulletin* **19**, 59–68.

PARK, P. B. 1975 Biodeterioration in aircraft fuel systems. In *Microbial Aspects of the Deterioration of Materials*, eds Lovelock, D. W. & Gilbert, R. J., pp. 105–126. London: Academic Press.

ROGERS, M. R. & KAPLAN, A. M. 1968 Screening of prospective biocides for hydrocarbon fuels. *Developments in Industrial Microbiology* **9**, 448–476.

ROGERS, M. R. & KAPLAN, A. M. 1982 Role of microbial and nonmicrobial contaminants in diesel-fuelled vehicle malfunctions. *Developments in Industrial Microbiology* **23**, 147–165.

SHERIDAN, J. E., NELSON, J. & TAN, Y. L. 1972 Studies on the 'kerosene fungus' *Cladosporium resinae* (Lindau) de Vries. Part 1. The problem of microbial contamination of aviation fuels. *Tuatara* **19**, 21–42.

SMITH, R. N. & CROOK, B. 1982 Biocides in fuels. *Biocides in the Oil Industry*, pp. 69–79. London: Institute of Petroleum.

Detection of Significant Spoilage Organisms in Soluble Oil-in-Water Metal-working Fluids

L. H. G. MORTON
School of Applied Biology, Lancashire Polytechnic, Corporation Street, Preston PR1 2TQ

Virtually all hydrocarbons are susceptible to attack and degradation by micro-organisms. For such an effect to occur water must be present; micro-organisms can only thrive in the aqueous phase associated with hydrocarbons (Traxler & Flannery 1968; Hill 1970; Genner & Hill 1981).

Oil-in-water emulsions have a wide and varied application in the engineering industry to cool and lubricate cutting and grinding operations. They are made by finely dispersing mineral oil in the presence of emulsifiers, usually with the addition of special additives to give extra lubricating film strength and anti-corrosive properties (Hill 1970). Many of these compounds present an array of degradable additives and nutrients.

Hydrocarbon-utilizing organisms have been shown to produce solubilizing or emulsifying extracellular agents which enable the organism to come into contact with the hydrocarbon droplets (Ratledge 1978). Many workers favour this theory of direct contact of cell with hydrocarbon droplet, rather than pre-solubilization of the hydrocarbon in the aqueous phase. The major utilization of the hydrocarbon component will only occur where there is a prolonged contact time between the hydrocarbon and the water phase, a situation which exists in machine tool coolant emulsions. The most obvious visible evidence of microbial growth in metal working fluid emulsions is the separation of the oil and aqueous phases. This results from chemical changes brought about by the activity of the organisms leading to an alteration in the oil droplet size distribution, producing an unstable emulsion (Garbutt 1973). Unless treated with an effective biocide, a metal-working fluid emulsion will invariably contain large numbers of microbial species (Bennett 1972). However, not all of these will be responsible for actual breakdown of fluid; only those able to metabolize one or more of the components (or breakdown products thereof) are considered as biodeteriogens. The test described below is that of Morton *et al.* (1983). It was devised in an attempt to recognize true biodeteriogens in the metal-working fluid microbial community.

Industrial
Microbiological Testing
0–632–01793–7

Materials Required

(1) A biocide-free mineral oil-based metal-working fluid. Suitable products for this test are Kutwell 40 which was supplied by Esso Research Ltd, Abingdon, and a formulation designated Soluble Oil A, supplied by Century Oils Ltd, Hanley. (2) Sterile distilled water. (3) Sterile glass bottles (medical flats) of 250 cm^3 capacity. (4) Nutrient agar and Nutrient broth (Oxoid Ltd). (5) Malt extract agar (Oxoid Ltd). (6) Tryptose blood agar base (Oxoid Ltd). (7) Bacteria/fungi isolated from in-service metal-working fluid(s). (8) Sterile pipettes (5 cm^3). (9) Sterile measuring cylinder 100 cm^3. (10) Sterilin bags.

Method

(i) *Preparation of metal-working fluid agars*
(a) 2.5 cm^3 of metal-working fluid is pipetted into 100 cm^3 of sterile distilled water in a sterile medical flat. The mixture is shaken thoroughly.
(b) 100 cm^3 of double-strength tryptose blood agar base and/or nutrient agar is prepared and sterilized by autoclaving. Whilst still molten, the agar is added to the metal-working fluid emulsion prepared as in (a). After mixing, 20 cm^3 aliquots of the metal-working fluid agar are poured into petri dishes and allowed to set. The plates should be dried prior to use.

(ii) *Isolation of micro-organisms from metal-working fluids*
Contaminated metal-working fluid is spread, using an inoculating loop, directly on to the surface of plates containing nutrient agar or malt agar. In the case of malt agar the addition of rose bengal has been found to suppress bacterial growth. The isolates should be brought into pure culture.

(iii) *Inoculation and incubation of metal-working fluids*
(a) Bacterial test organisms are cultivated in nutrient broth for 24–36 h and are either spotted or streaked onto the surface of dried metal-working fluid agar plates. Incubation is at 32°C for up to 14 days.
(b) Fungal spores or mycelial fragments are inoculated directly onto the surface of metal-working fluid agar. These plates are incubated at 25°C for up to 14 days.

All plates should be placed in sealed 'Sterilin' bags containing moist cotton wool in order to maintain a high humidity during the incubation period.

Interpretation of Results

Fig. 1 shows the distinct clearing zones produced in Soluble Oil A by *Pseudomonas aeruginosa*. Fig. 2 shows clearing produced by various other bacteria isolated from

FIG. 1. Clearing of Soluble Oil A by *Pseudomonas aeruginosa*.

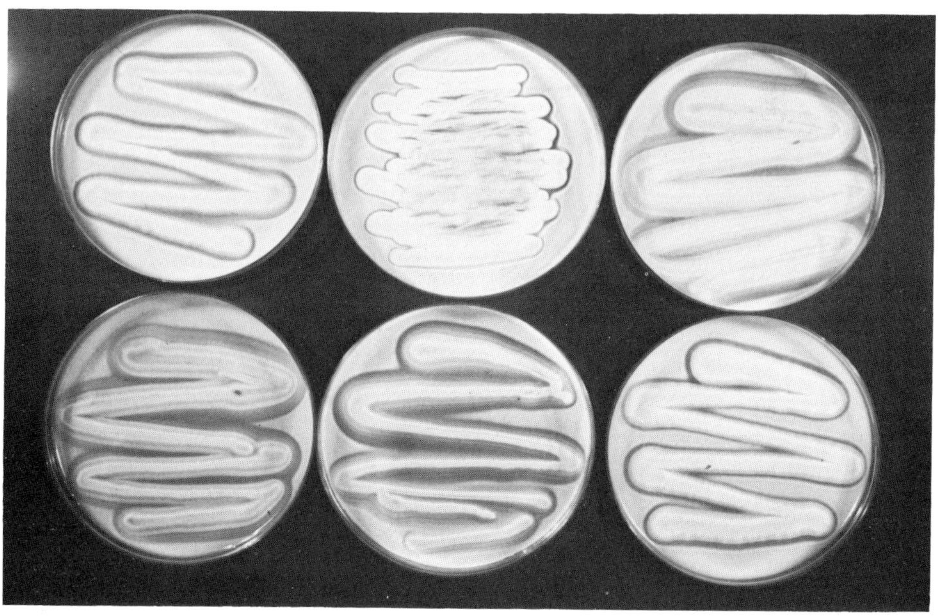

FIG. 2. Clearing produced by various bacteria isolated from contaminated metal-working fluids.

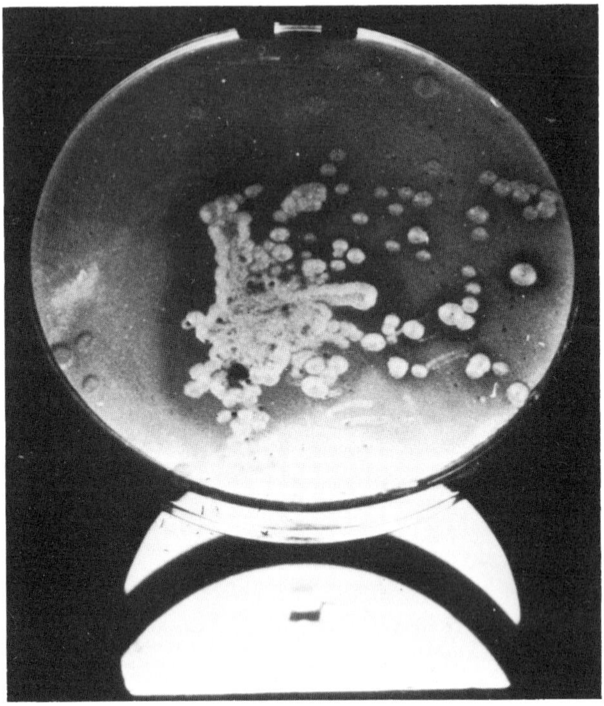

FIG. 3. Clearing of Kutwell 40 by *Aspergillus flavus*.

metal-working fluids. Fig. 3 shows the clearing produced by *Aspergillus flavus* when grown on Kutwell 40. A list of organisms found to clear metal-working fluid agar is recorded in Morton *et al.* (1983). Whether the clearing of the agar is brought about by the enzymic breakdown of the oil droplets or whether coalescence of droplets results from an emulsion instability effect is still under consideration. A report of collaborative experiments undertaken by members of the metal-working fluid group of the IBRG (International Biodeterioration Research Group) points to the clearing being due to the production of some extracellular product(s) (Morton 1985).

References

BENNETT, E. O. 1972 The biology of metal-working fluids. *Lubrication Engineering* **28**, **(6)**, 237–247.

GARBUTT, J. 1973 Microbes can eat into cooling efficiency. *Metal-working Production* **117**, 126–129.

GENNER, C. & HILL, E. C. 1981 Fuels and Oils. *Microbial Biodeterioration*, ed. Rose, A. H., pp. 260–306. London: Academic Press.

HILL, E. C. 1970 The microbiology of industrial lubricants and emulsions. *Medicine* **39**, **(2)**, 47–50.

MORTON, L. H. G. 1985 A report on the IBRG collaborative experiment 'An examination of the clearing of a metal-working fluid agar as a means of detecting bacteria which degrade metal-working fluids'. IBRG Document No. MWF/84/09.

MORTON, L. H. G., GILLATT, J. W., WARRILOW, E. O. & GREENHALGH, M. 1983 A potential method for the recognition of metal-working fluid spoilage organisms. *International Biodeterioration Bulletin* **19**, (2), 39–43.

RATLEDGE, C. 1978 Degradation of aliphatic hydrocarbons. *Developments in Biodegradation of Hydrocarbons—1*, ed. Watkinson, R. J., pp. 1–47. London: Applied Science Publishers.

TRAXLER, R. W. & FLANNERY, W. L. 1968 Mechanisms of hydrocarbon degradation. In: *Biodeterioration of Materials*, eds Walters, A. J. & Elphick, J. H. London: Elsevier.

A Recirculating Test Rig for the Investigation of Metal-working Fluid Spoilage

A. P. RAWLINSON AND JEAN L. SHENNAN
BP Research Centre, Sunbury-on-Thames

If control measures are not taken, soluble oil emulsions used for cooling, lubrication and swarf removal in metal machining operations are prone to premature product wastage following microbial contamination.

Chronic contamination of a machine system using unprotected cutting fluids can easily build up, shortening the effective life of a given product from 3 or more months to a matter of days. If uncontrolled, microbial contamination can reach counts of 10^8–10^9/ml in the circulating fluid and may lead to the formation of surface slimes, sludges associated with swarf accumulation or floating mats of fungal mycelia. Although such aggregations of biomass can be partly removed by swarf separators or filters, heavy contamination usually leads quickly to emulsion breakdown, especially if the emulsifier has been used as a growth substrate. However, some soluble oil emulsions can sustain a high bacterial count without loss of performance. In these cases, the 'Monday morning smell' of hydrogen sulphide, disengaged from an anaerobic swarf-laden sump when the fluid circulation system is turned on after a weekend shutdown, will call for action if no other indication of spoilage has appeared. But for the presence of free H_2S, these products would continue to be acceptable as lubricoolants.

Contamination control by the addition of antimicrobial chemicals, biocides, is the short-term solution to the problem of emulsion spoilage (Hill *et al.* 1976; Rossmoore 1981; Shennan 1983). In the longer term the answer is to formulate new soluble oil emulsions using ingredients with greater intrinsic resistance to biodegradation or completely new coolants which are microbially less susceptible.

An intermediate target for control is to reduce the problem to manageable proportions. For this, some understanding is required of the microbial ecosystem represented by a soluble oil emulsion undergoing active biodeterioration. Many factors other than the degradability of the formulation ingredients can contribute to cutting fluid spoilage—swarf accumulation, atmospheric oxidation, exhaustion of active chemicals, make-up water quality and general workshop practices. Keeping

Industrial
Microbiological Testing
0–632–01793–7

contamination under control, rather than attempting elimination of micro-organisms, may be a practical aim.

Published methods for investigating the spoilage of cutting fluids, and for the selection and testing of suitable biocides, have been reviewed by Shennan (1983). These methods fall into two main categories: (i) those using large aerated culture bottles (e.g. Izzat & Bennett 1979; Rossmoore *et al.* 1979), and (ii) those using a recirculating system (Pivnick & Fabian 1953; Himmelfarb & Scott 1968; Hill *et al.* 1976).

Recirculation rig methods attempt to simulate the conditions under which a cutting fluid will be used in service. The test rig described in this paper has been developed over a number of years in the authors' laboratory and used extensively for the testing of new product formulations and the evaluation of biocides.

The Method

Recirculating test rig

The basic unit (Fig. 1) consists of a 1 litre tall glass beaker containing 800 ml of test fluid and 2 g of cast iron chips. The beaker cover carries a glass funnel containing a further 2 g of mild steel millings supported on metal gauze. The inlet air flow enters the base of the unit through a specially blown glass tube and forces fluid up the flow tube to discharge without splashing over the swarf in the funnel to trickle back into the reservoir. The air flow is regulated to give a fluid circulation of *c.* 5 litres per hour and is turned off overnight and at weekends.

A battery of such beakers, allowing several treatments with replication to be assessed in one test run, is held in a water bath at *c.* 30°C (Fig. 2). The water bath heating is turned off at weekends. The entire recirculating rig is set up in a fume cupboard to minimize any potential hazard from oil mists or aerosols.

Each beaker is inoculated with 5 ml of a mixed microbial population prepared from isolated cutting fluid contaminants maintained in a biocide-free soluble oil emulsion together with samples of fresh spoiled fluid. Fluids of a similar type to those being screened are chosen, where possible, for inoculum preparation. The initial count in the test fluid is 10^5–10^6/ml bacteria and *c.* 10^3/ml yeasts and moulds.

Samples (5 ml) are withdrawn for microbial counts as required, once or twice weekly, after making up the fluid level with distilled water. At the same time pH is measured and the emulsion strength of the fluid estimated with a pocket refractometer. At the start-up of aeration each Monday, tests for liberation of hydrogen sulphide gas, using lead acetate-impregnated paper, are carried out on individual units. If required, for instance if a fluid formulation shows biocidal activity and reduces the viable microbial population, the units can be reinoculated with a fresh culture equivalent to the initial challenge at suitable intervals.

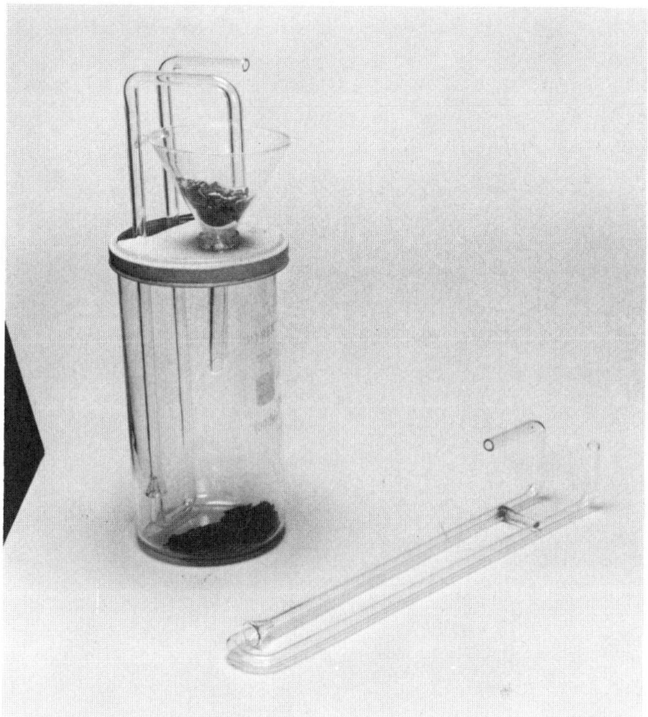

FIG. 1. Basic unit of the recirculating test rig.

Enumeration methods

Standard techniques were used to estimate viable cell numbers following serial dilution in Ringer's peptone saline solution if necessary. For example, total aerobic bacteria can be enumerated on Plate Count Agar (e.g. Oxoid CM 325) and yeasts and moulds on Sabouraud Dextrose Agar (e.g. Oxoid CM 41) with 0.001% chlortetracycline. Peptone cystine sulphate broth (Skerman 1967) can be used to estimate sulphide-generating bacteria on a presence-or-absence basis in a given aliquot.

Discussion

Laboratory recirculating tests for investigating aqueous metal-working fluids are favoured by many workers as they simulate the conditions under which a cutting fluid will be used in service. In all versions the emulsions are circulated over or through metal chips. The test systems are non-aseptic and are subjected to repeated

FIG. 2. Recirculating rig for cutting fluid testing.

challenges using a mixed flora of contaminants, with continuous or interrupted aeration. The detailed design of the systems also allows differing degrees of aerobiosis to develop, simulating the aerated metal surface of the workpiece and the anaerobic conditions in a swarf-laden sump.

No universally suitable inoculum can be devised for all types of cutting fluid investigations. A defined inoculum may be useful for certain types of comparative screening test, but the cutting fluid formulator is ultimately concerned with the performance of the product under machine shop conditions. The in-service conditions of continual and mixed microbial challenge are best simulated by an inoculum of spoiled fluid taken from the working environment or a mixture of organisms recently isolated from and maintained in similar fluids. To assist comparison between test runs, it is recommended that a positive control, i.e. a fluid of known susceptibility, be included in each batch of fluids tested.

The recirculation pattern illustrated above is suitable for soluble oil emulsions based on mineral oils and used in machine shops working on a 5-day week basis. By adjustment of the apparatus and the operational regime, a variety of metal-working fluid types and in-service usage patterns can be simulated.

The proof of simulated procedures lies in the comparison of results obtained in screening tests with those obtained from subsequent field trials. The method described above has been proven over a number of years of cutting fluid product development and biocide screening tests.

Acknowledgments

Many colleagues at the BP Research Centre have been involved in the development of the recirculating rig described in this paper. The authors would like to acknowledge their contributions and to thank British Petroleum plc for permission to publish this paper.

References

HILL, E. C., GIBBON, O. & DAVIES, P. 1976 Biocides for use in oil emulsions. *Tribology International* **9**, **(3)**, 121–130.

HIMMELFARB, P. & SCOTT, A. 1968 Simple circulating tank test for evaluation of germicides for cutting fluid emulsions. *Applied Microbiology* **16**, 1437–1438.

IZZATT, I. N. & BENNETT, E. O. 1979 The potentiation of cutting fluid preservatives by diethylene triamine pentacetic acid. *International Biodeterioration Bulletin* **15**, **(1)**, 1–6.

PIVNICK, H. & FABIAN, F. W. 1953 Methods for testing the germicidal value of chemical compounds for disinfecting soluble oil emulsions. *Applied Microbiology* **1**, 204–207.

ROSSMOORE, H. W. 1981 Antimicrobial agents for water-based metal-working fluids. *Journal of Occupational Medicine* **23**, **(4)**, 247–254.

ROSSMOORE, H. W., SIECKHAUS, J. F., ROSSMOORE, L. A. & DEFONZO, D. 1979 The utility of biocide combinations in controlling mixed microbial populations in metalworking fluids. *Lubrication Engineering* **35**, 559–563.

SHENNAN, J. L. 1983 Selection and evaluation of biocides for aqueous metal-working fluids. *Tribology International* **16**, **(16)**, 317–330.

SKERMAN, V. B. D. 1967 *A Guide to the Identification of the Genera of Bacteria.* Baltimore: Williams & Wilkins.

The Bridge Slide Culture Techniques for Fungicide Studies in Fuel Oil

R. N. SMITH AND J. M. MAW
Microbiology Group, The Hatfield Polytechnic, Hatfield, Herts., U.K.

A wide range of bacteria, yeasts and filamentous fungi are able to utilize hydrocarbons as carbon and energy sources (Ratledge 1978). *Hormoconis resinae* (*Cladosporium resinae*) is a major deteriogen of C9–C19 fuel oils and has been widely reported as a deteriogen of jet fuels and in fuel oil for marine and terrestrial gas turbine engines.

Fungal growth on fuel oil occurs almost exclusively at the oil–water interface. The very low solubility of n–paraffins greater than C10 in water is such that dissolved hydrocarbon can be discounted as a nutrient source (Miura 1978). Thus the behaviour of such fungi can only be understood by observations made at the oil–water interface. In favourable conditions unattached hyphal growth at the oil–water interface can rapidly develop into a thick mycelial mat. Hyphal growth by *H. resinae* has also been shown to develop in small puddles of water where it ramifies over the oil–water interface thereby anchoring the water and becoming a site for further water entrapment (Elphick & Hunter 1968).

The two methods described below were designed to enable quantitative determination of the effects of fuel oils containing antifungal materials on fungal deteriogens at the oil–water interface.

In one the behaviour and rate of germination of conidia of *H. resinae* suspended in water under fuel oil is recorded. The relationship between the probit value of percentage inhibition and the logarithm of fungicide concentration (Wilcoxon & McCallan 1939; Bliss 1957) can then be established.

In the other, conidia of *H. resinae* are germinated on an agar platform under fuel oil containing sublethal concentrations of candidate fungicides and the specific growth rate of the vegetative hyphae compared with the rate obtained under untreated fuel oil. Trinci (1974) has shown that during the first 12 h of growth of a fungal colony arising from a single spore, the increase in total hyphal length (equivalent to biomass) is exponential. When the natural logarithm of total hyphal length is plotted against time a linear relationship is obtained in which the slope of

Industrial Microbiological Testing
0–632–01793–7

the line represents the specific growth rate. Thus the effect of a fungicide on the specific growth rate can be determined, and by relating this to the rate of hyphal elongation, the effect on the size of the peripheral zone can also be determined.

Materials and Equipment

Construction of bridge slides

Bridge slides are made from standard microscope slides which have been modified to support a large rectangular coverslip 500 μm above the surface of the slide. Two stainless steel coupons 20 mm × 10 mm × 500 μm are fixed 30 mm apart across a clean slide using a polysulphide rubber adhesive (PR1422ZBT 1/2 Berger Chemical Ltd).

Culture media

Malt extract agar. Malt extract agar was prepared from a dehydrated mixture (Oxoid CM 60) which contained (g/l): malt extract, 30 g; mycological peptone, 5 g. The pH was adjusted to 5.4 and the medium sterilized by autoclaving for 10 min at 115°C.

Modified Bushnell & Haas medium. Bushnell and Haas medium as modified by Parberry & Thistlethwaite (1973) was prepared as follows: $MgSO_4.7H_2O$, 0.02 g; $CaCl_2.2H_2O$, 0.002 g; K_2HPO_4, 0.1 g; NH_4NO_3, 0.95 g; $FeCl_3$, 2 drops of a 1/10 dilution of a saturated solution; distilled water, 1 litre; agar, 20.0 g was added where necessary. The pH was adjusted to 6.0 and the solution autoclaved for 15 min at 121°C.

Preparation of conidial suspension

Conidial suspensions of *H. resinae* are prepared from cultures grown on malt extract agar slopes at 25°C. Sterile quarter-strength Ringer's solution containing 0.1% Tween 80 is shaken over the slopes and the resulting suspension filtered through sterile muslin or three layers of nylon mesh (pore size 106 μm) to remove hyphal fragments, centrifuged and resuspended in modified Bushnell & Haas liquid medium.

Preparation of test solutions of fungicide in oil

25 mg fungicide is dissolved in 0.5 ml dimethyl sulphoxide (DMSO). This is then shaken with 1 ml fuel oil and the resulting mixture made up to 50 ml in additional fuel oil to give a stock preparation containing 500 ppm (w/v) of the test compound. These stock solutions can be held in a refrigerator and used to prepare test solutions of fungicide in fuel oil. All fuel oil samples are sterilized by filtration before use.

Test Procedures

Use of the bridge slide for spore germination studies

A 30 μl drop of modified Bushnell & Haas medium containing 250 000 conidia of *H. resinae* per ml is placed in the centre of the slide between the steel coupons and spread over an area 10 × 10 mm to give a rectangular drop 300 μm deep. A coverslip 40 × 22 mm is then placed across the steel coupons and fuel oil is run in under the coverslip using a Pasteur pipette. In practice it is advisable to moisten the underside of the coverslip with oil before placing on the slide to prevent water from the drop condensing on the coverslip prior to flooding with oil. The volume of oil under each slide is approximately 300 μl, giving an oil:water ratio of 10:1 (Fig. 1).

Use of the bridge slide to determine the specific growth rate of vegetative hyphae

Agar slide cultures under oil are achieved by placing a thin platform of modified Bushnell & Haas agar on the bridge slide. The agar surface is inoculated with conidia of *H. resinae*, covered with a coverslip and the remaining space flooded with fuel oil.

The agar platform is prepared as follows. A microscope slide is prepared bearing at its centre two coverslips 18 × 18 mm, each 100 μm thick, glued one above the other so that it has a raised area 200 μm high. A small volume (0.75 ml) of molten agar at 50°C is placed on a bridge slide between the stainless steel coupons. The slide bearing the two coverslips is laid firmly across the two steel coupons of the bridge slide with the coverslips on the underside. When the agar has set the top slide is carefully removed leaving a moulded layer of agar 500 μm thick containing a depression, 18 mm square and 200 μm deep, between the steel coupons. The agar surrounding the square depression is cut away leaving a platform of 10 mm square and 300 μm thick.

This agar platform is inoculated with 10 μl of a conidial suspension of *H. resinae* containing 1000 conidia per ml and spread carefully over the surface of the agar. The agar bridge slide is left for 10 min to allow the agar to absorb the free water. A

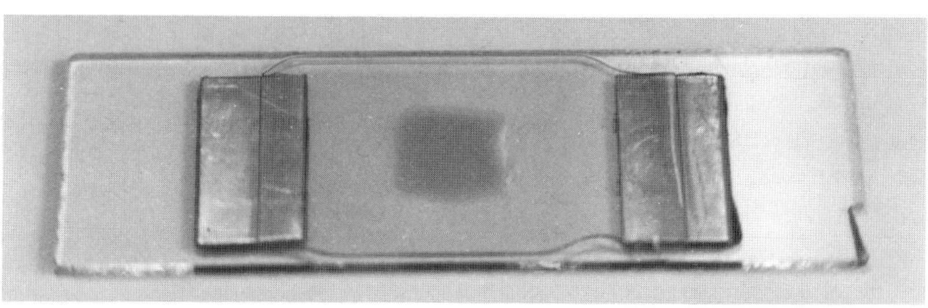

FIG. 1. Bridge slide with a 1 cm square of modified Bushnell & Haas liquid medium containing *H. resinae* covered by oil under a coverslip 40 × 22 mm.

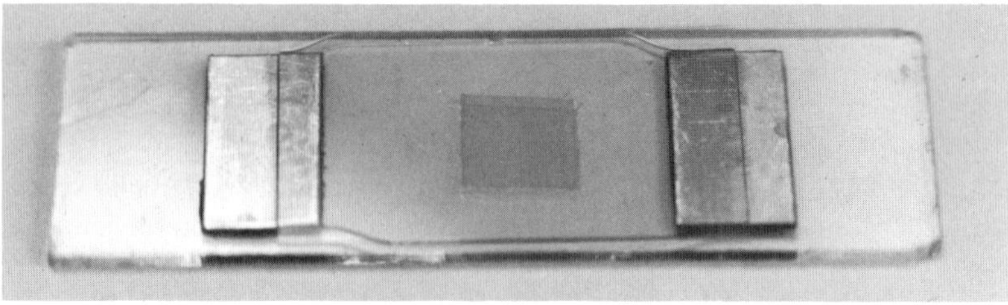

FIG. 2. Agar bridge slide with a platform of agar 10 × 10 × 0.5 mm in the centre, covered by oil under a coverslip 40 × 22 mm.

40 × 22 mm No. 1 coverslip is then placed across the steel coupons and fuel oil is run under the coverslip using a sterile Pasteur pipette (Fig. 2). The agar slide cultures are placed in covered dishes and incubated at 25°C. After 14 h incubation the slides are examined under the microscope. Single conidia which have commenced growth are selected and their position recorded using the vernier scales on the microscope stage. The agar surface is smooth and permits fungal conidia and hyphae to be observed with ease. Any germinating spores are relocated at set time intervals and either photographed or the outline of each hyphae quickly traced on paper using a drawing tube. Both methods permit hyphal development to be rapidly recorded and several colonies on one slide can be examined in a few minutes. The total hyphal lengths which have developed at each time can be determined from photographs or line drawing using a map measurer.

Results and Discussion

Use of the bridge slide for spore germination studies

When spore suspensions were covered by untreated oil, most of the spores germinated within 3 days, and by the fourth day an interwoven mycelium had developed in which no individual colonies could be distinguished. The behaviour of the hyphae emerging from individual conidia seemed to be influenced by their proximity to the oil–water interface. Those hyphae which were in suspension in the aqueous phase often developed into a short conidiophore with a small cluster of conidia soon after germination. In contrast, conidia which germinated close to the oil–water interface frequently produced long vegetative ‘feeding’ hyphae which grew out into the oil (Fig. 3). Colonies formed in this way developed an extensive mycelium along the oil–water interface with hyphae penetrating some distance into the oil (Fig. 4). The hyphae growing out into the oil appeared to secrete small droplets of water along their length (Fig. 5). Conidiophores were produced from hyphae which grew down from the mycelium at the oil–water interface into the

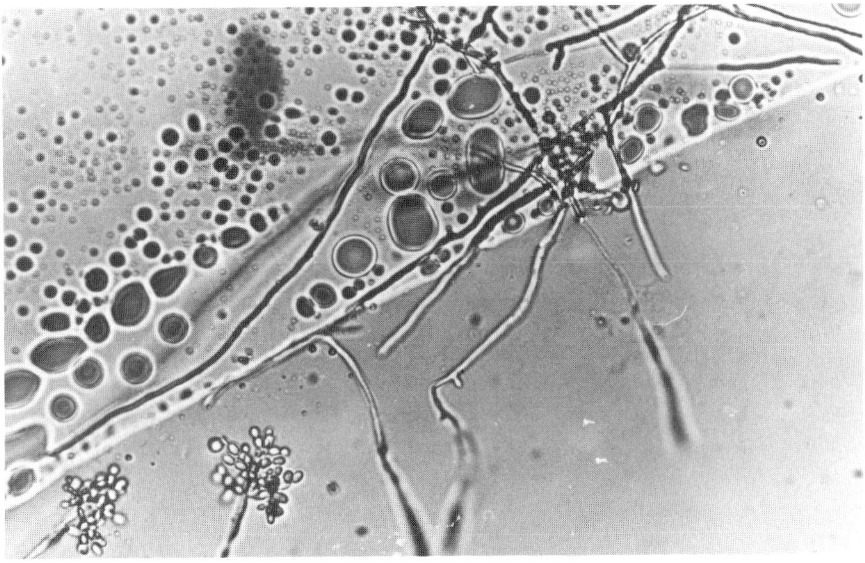

FIG. 3. Interface at the edge of a Bushnell & Haas medium droplet showing long feeding hyphae growing into the oil phase and conidiophores developing in the aqueous phase. The oil phase is in the top left and the water phase is the lower right-hand portion of the figure.

aqueous phase where the conidia were formed (Fig. 4). The restriction of extensive hyphal growth to the oil–water interface may be a nutritional effect but may also be due to the high solubility of oxygen in the oil. Teh & Lee (1973) reported that in static culture maximal growth of *H. resinae* in a two-phase oil and water system only occurred when the oil:water ratio was greater than 1:9.

Conidia under oil with antifungal properties showed two symptoms: the first was a failure to germinate, and the second was the production of short bent or deformed germ tubes which did not grow any further (Fig. 6). These could be contrasted with the long branching vegetative hyphae and conidiophores of *H. resinae* in the control treatments. In oil containing carbendazim, normal germination and development of vegetative hyphae only occurred when the concentration of carbendazim was less than 1.0 ppm (Smith & Crook 1980). Conidia germinated under oil containing 1.0, 3.0 and 10.0 ppm carbendazim, but they produced only short deformed germ tubes and further growth was arrested. The results showed that the suppression of germination was related to the concentration of carbendazim. These rates have been expressed as percentage inhibition compared with the control and their corresponding probit value are given in Table 1.

Use of the bridge slide to determine the specific growth rate of vegetative hyphae

Young developing colonies of *H. resinae* were clearly visible and were easily

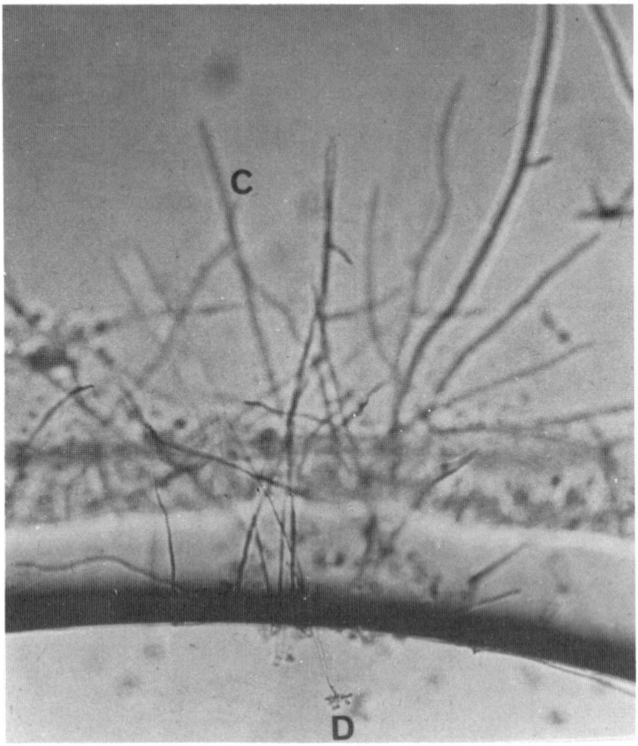

FIG. 4. The oil–water interface taken from the edge of a Bushnell & Haas droplet showing extensive nonsporing feeding hyphae growing out into the oil (C) and conidiophores in the aqueous phase (D).

photographed in agar slide culture. These colonies were held on the surface of the agar at the oil–water interface and it was found that very few of the young hyphae were converted to conidiophores, supporting an earlier observation that close contact with fuel oil encourages vegetative growth. Measurements of total hyphal

TABLE 1. *Effect of carbendazim in oil on the germination rate of conidia of* H. resinae *over the first 42 h*

Concentration			
ppm	log	% Germination	Probit % inhibition
0.0	—	76	—
0.3	0.5	61	4.16
1.0	0.0	52	4.48
3.0	0.5	30	5.25
10.0	1.0	8	6.28

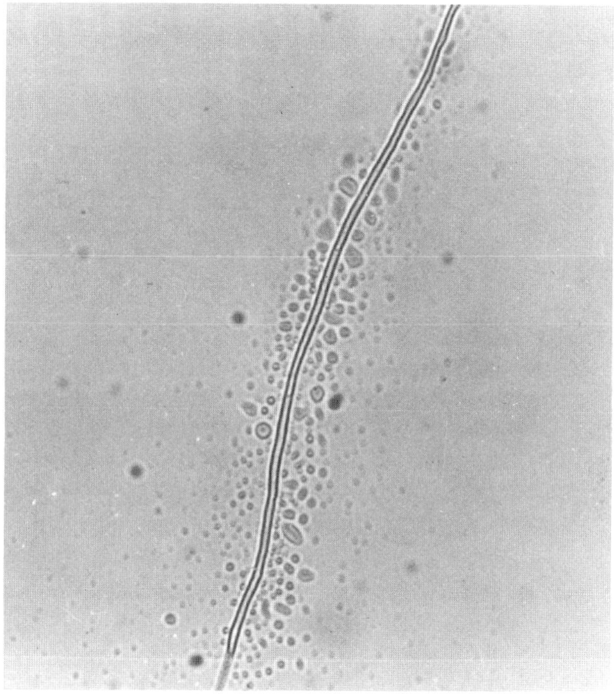

FIG. 5. Feeding hyphae of *H. resinae* surrounded by water droplets excreted from the hypha (× 400).

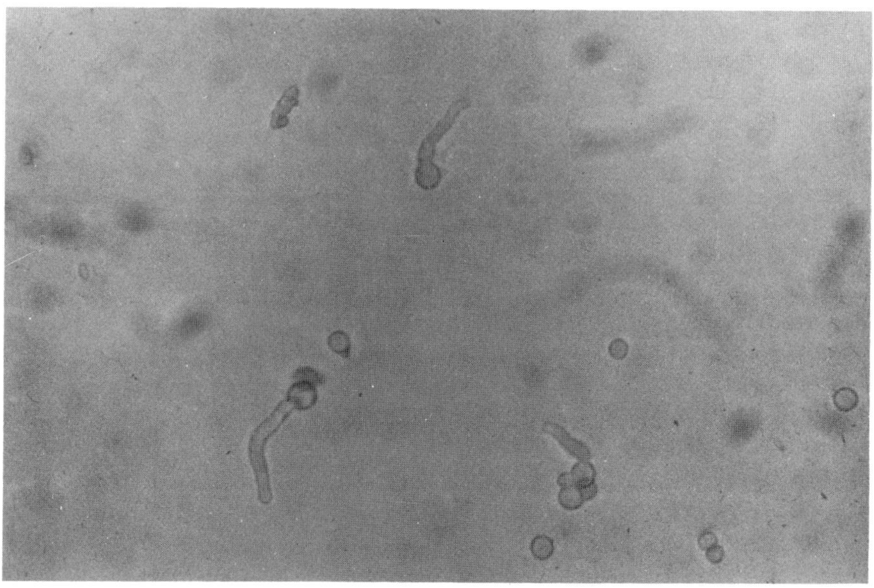

FIG. 6. Conidia of *H. resinae* germinating under oil containing 3 ppm carbendazim. Germ tubes are severely distorted and growth has ceased.

TABLE 2. *Specific growth rates of* H. resinae *on malt extract agar and modified Bushnell & Haas agar under fuel oil*

Treatment	Specific growth rate per hour
ME + oil	0.26
B & H + oil	0.23
B & H + oil + 0.3 ppm carbendazim	0.21
B & H + oil + 1.0 ppm mildothane	0.18
B & H + oil + 1.0 ppm thiram	0.20

length were made, the natural logarithms of these values were plotted and the specific growth rate for each treatment is given in Table 2.

These slides may also be used to determine the size of the hyphal growth unit. The combined results could be used to construct a model to predict the extent of hyphal growth at the oil–water interface under specified conditions.

References

BLISS, C. I. 1957 Some principles of bioassay. *American Scientist* **45**, 449–466.

ELPHICK, J. J. & HUNTER, S. K. P. 1968 Evaluating biocidal fuel additives for intermittent use in aircraft fuel systems. *Biodeterioration of Materials*, Vol. 1, eds Walters, A. H. & Elphick, J. J., pp. 364–370. London: Applied Science Publishers.

MIURA, Y. 1978 Mechanism of liquid hydrocarbon uptake by microorganisms and growth kinetics. *Advances in Biochemical Engineering* **9**, 31–36.

PARBERRY, D. G. & THISTLETHWAITE, P. J. 1973 Improved mineral salts solution for the growth of *Amorphotheca resinae*. *International Biodeterioration Bulletin* **9**, 11–16.

RATLEDGE, C. 1978 Degradation of aliphatic hydrocarbons. *Developments of Biodegradation of Hydrocarbons—1*, ed. Watkinson, R. J., pp. 1–46. London: Applied Science Publishers.

SMITH, R. N. & CROOK, B. 1980 The germination and growth of *Cladosporium resinae* in fuel. *Biodeterioration*, eds Oxley, T. A., Allsop, D. & Becker, G., pp. 29–36. London: Pitman Publishing.

TEH, J. S. & LEE, K. H. 1974 Effects of n-alkanes on *Cladosporium resinae*. *Canadian Journal of Microbiology* **20**, 971–976.

TRINCI, A. P. J. 1974 A study of the kinetics of hyphal extension and branch initiation of fungal mycelia. *Journal of General Microbiology* **81**, 225–236.

WILCOXON, F. & MCCALLAN, S. E. A. 1939 Theoretical principles underlying laboratory toxicity tests of fungicides. *Contributions of the Boyce Thompson Institute* **10**, 329–338.

Index

THE SOCIETY FOR APPLIED BACTERIOLOGY TECHNICAL SERIES

General Editor: F. A. Skinner

1 Identification Methods for Microbiologists Part A *1966*
Edited by B. M. Gibbs and F. A. Skinner
Out of print

2 Identification Methods for Microbiologists Part B *1968*
Edited by B. M. Gibbs and D. A. Shapton
Out of print

3 Isolation Methods for Microbiologists *1969*
Edited by D. A. Shapton and G. W. Gould

4 Automation, Mechanization and Data Handling in Microbiology *1970*
Edited by Ann Baillie and R. J. Gilbert

5 Isolation of Anaerobes *1971*
Edited by D. A. Shapton and R. G. Board

6 Safety in Microbiology *1972*
Edited by D. A. Shapton and R. G. Board

7 Sampling—Microbiological Monitoring of Environments *1973*
Edited by R. G. Board and D. W. Lovelock

8 Some Methods for Microbiological Assay *1975*
Edited by R. G. Board and D. W. Lovelock

9 Microbial Aspects of the Deterioration of Materials *1975*
Edited by R. J. Gilbert and D. W. Lovelock

10 Microbial Ultrastructure: The Use of the Elecron Microscope *1976*
Edited by R. Fuller and D. W. Lovelock

11 Techniques for the Study of Mixed Populations *1978*
Edited by D. W. Lovelock and R. Davies

12 Plant Pathogens *1979*
Edited by D. W. Lovelock

13 Cold Tolerant Microbes in Spoilage and the Environment *1979*
Edited by A. D. Russell and R. Fuller

14 Identification Methods for Microbiologists (2nd Edn) *1979*
Edited by F. A. Skinner and D. W. Lovelock

15 Microbial Growth and Survival in Extremes of Environment *1980*
Edited by G. W. Gould and Janet E. L. Corry

16 Disinfectants: Their Use and Evaluation of Effectiveness *1981*
Edited by C. H. Collins, M. C. Allwood, Sally F. Bloomfield and A. Fox

17 Isolation and Identification Methods for Food Poisoning Organisms *1982*
Edited by Janet E. L. Corry, Diane Roberts and F. A. Skinner

18 Antibiotics: Assessment of Antimicrobial Activity and Resistance *1983*
Edited by A. D. Russell and L. B. Quesnel

19 Microbiological Methods for Environmental Biotechnology *1984*
Edited by J. J. Grainger and J. M. Lynch

20 Chemical Methods in Bacterial Systematics *1985*
Edited by M. Goodfellow and D. E. Minnikin

21 Isolation and Identification of Micro-organisms of Medical and Veterinary Importance *1985*
Edited by C. H. Collins and J. M. Grange

22 Preservatives in the Food, Pharmaceutical and Environmental Industries *1986*
Edited by R. G. Board, M. C. Allwood and J. G. Banks

23 Industrial Microbiological Testing *1987*
Edited by J. W. Hopton and E. C. Hill

24 Immunological Techniques in Microbiology *1987*
Edited by J. M. Grange, A. Fox and N. L. Morgan